Lecture Notes in Chemistry

Edited by G. Berthier, M. J. S. Dewar, H. Fischer
K. Fukui, H. Hartmann, H. H. Jaffé, J. Jortner
W. Kutzelnigg, K. Ruedenberg, E. Scrocco, W. Zeil

16

W0036583

Petr Čársky
Miroslav Urban

Ab Initio Calculations
Methods and Applications in Chemistry

Springer-Verlag
Berlin Heidelberg New York 1980

Authors

Petr Čársky
J. Heyrovsky Institute of Physical
Chemistry and Electrochemistry
Czechoslovak Academy of Science
12138 Prague 2
Czechoslovakia

Miroslav Urban
Department of Physical Chemistry
Faculty of Science
Comenius University
81631 Bratislava
Czechoslovakia

ISBN-13: 978-3-540-10005-8   e-ISBN-13: 978-3-642-93140-6
DOI: 10.1007/978-3-642-93140-6

2152/3140-543210

Preface

Until recently quantum chemical ab initio calculations were restricted to atoms and very small molecules. As late as in 1960 Allen
and Karo stated[1]: "Almost all of our ab initio experience derives
from diatomic LCAO calculations ..." and we have found in the literature "approximately eighty calculations, three-fourths of which are
for diatomic molecules ... There are approximately twenty ab initio
calculations for molecules with more than two atoms, but there is a
decided dividing line between the existing diatomic and polyatomic
wave functions. Confidence in the satisfactory evaluation of the many-
-center two-electron integrals is very much less than for the diatomic case". Among the noted twenty calculations, SiH_4 was the largest
molecule treated. In most cases a minimal basis set was used and the
many-center two-electron integrals were calculated in an approximate
way. Under these circumstances the ab initio calculations could hardly provide useful chemical information. It is therefore no wonder
that the dominating role in the field of chemical applications was
played by semiempirical and empirical methods. The situation changed
essentially in the next decade. The problem of many-center integrals
was solved, efficient and sophisticated computer programs were developed, basis sets suitable for a given type of problem were suggested,
and, meanwhile, a considerable amount of results has been accumulated
which serve as a valuable comparative material. The progress was of
course inseparable from the development and availability of computers.
Further progress was achieved in the early seventies' by the development of procedures that yield correlated wave functions for small molecules at a cost moderately higher than that involved in SCF calculations.

Existing books on this subject are mostly textbooks for advanced
readers, proceedings of conferences, or edited collections of reviews
written by several authors on selected special topics. These are mainly profitable for readers who already have some experience with the
ab initio calculations. Our approach is different. We attempted a book
which would be a practical handbook for those who are going to start
their ab initio calculations. It is hoped that it permits answers to
be found readily to the most common questions put to us by students
and colleagues from other branches of theoretical chemistry. Ideally,
our aim was to write such a book which would have been helpful for us
in our first own ab initio calculations. Our intention is to deal solely with the problems specific for ab initio calculations. We expect

that the reader is familiar with the fundamentals of the molecular orbital theory and that he already has some experience with semiempirical methods or at least with the Hückel and extended-Hückel methods.
We emphasize the following topics: the basis set, the correlation energy, their effects on results in actual chemical applications and the
relation between the cost and the use of ab initio calculations. The
choice of the topics treated and the coverage are of course arbitrary
and they were affected both by personal taste and the limited knowledge.

The writing of this book was stimulated by the suggestion of Dr.
Rudolf Zahradník, Dr.Sc. His encouragement, valuable comments and as-
sistance of various kinds are gratefully acknowledged. We are also
much obliged to Dr. Ivan Hubač, who made us familiar with the theories
and methods of calculation of the correlation energy and made available to us all his unpublished material on this topic. Our sincere
thanks are also due to Dr. Pavel Hobza for comments on the sections
on weak interactions and solvation and Dr. Jiří Pancíř for comments
on the geometry optimization methods. For typing the manuscript and
drawing the figures we wish to acknowledge our appreciation to Mrs.
Růžena Žohová and Mrs. Erika Týleová.

Prague and Bratislava, June 1979

P.Č. and M.U.

Contents

1. Introduction

For a nonspecialist the term "ab initio" may have a too authoritative meaning. It might give the impression that the ab initio calculations provide "accurate" and "objective" solutions of the Schrödinger equation. In fact, they can be in considerable, sometimes even in qualitative, disagreement with experimental findings. "Ab initio" only implies that within the frame of a particular variation or perturbation method no approximations are adopted, though the method itself is a mere approximation to the solution of the Schrödinger equation. That means that unlike in semiempirical methods no integrals are neglected or approximated by simplified expressions and functions containing empirical parameters, or even replaced by empirical parameters. Explicit inclusion is also made for inner shell electrons. All integrals should be calculated with an accuracy higher than to 7-8 digits. "Ab initio" also implies that a nonrelativistic Hamiltonian within the Born-Oppenheimer approximation is used

$$ H = -\frac{1}{2}\sum_i \nabla_i^2 - \sum_{i,A} \frac{Z_A}{r_{iA}} + \sum_{i>j} \frac{1}{r_{ij}} + \sum_{A>B} \frac{Z_A Z_B}{R_{AB}} \tag{1.1} $$

Indices i,j refer to electrons, indices A,B to nuclei and Z_A,Z_B are nuclear charges. Hamiltonian (1.1) is expressed in "atomic units". If not otherwise stated, atomic units will be used throughout (see Appendix A). Hence the ab initio calculations do not involve any experimental data. They are therefore also referred to as "nonempirical" calculations.

The definition of ab initio calculations given above corresponds to the definition by Allen and Karo[1] from 1960. As we shall see in Chapter 3, it does not conform perfectly to the present practice as regards the requirements noted above. Nowadays a justifiable neglect is commonly made for almost vanishing integrals which does not affect the computed total energy. But the term ab initio is also used for calculations that are lacking this rigor. These are for example nonempirical calculations using pseudopotentials or even "simplified" and "simulated" ab initio calculations in which much cruder approximations are used.

In this book we shall mainly treat calculations obeying the traditional definition. In order to judge the possibilities of the ab

T a b l e 1.1

Selected ground state atomic characteristics[a] obtained by the
Dirac-Breit-Pauli-Hartree-Fock method[2,3]

Atom	Hartree-Fock energy	Normal and specific mass effects	Total relativistic correction
He	-2.861697	0.00039	-0.0000708
Li	-7.432729	0.000583	-0.0005504
Be	-14.57303	0.000888	-0.0021914
B	-24.52906	0.00120	-0.0061116
C	-37.68866	0.00181	-0.0138302
N	-54.40098	0.00201	-0.0270472
O	-74.80947	0.00223	-0.0490197
F	-99.40944	0.00259	-0.0820816
Ne	-128.5472	0.00297	-0.1289864
Na	-161.8591	0.00337	-0.1968539
Mg	-199.6145	0.00378	-0.2899070
Al	-241.8768	0.00420	-0.4136049
Si	-288.8544	0.00463	-0.5740692
P	-340.7188	0.00507	-0.7773523
S	-397.5050	0.00551	-1.033528
Cl	-459.4822	0.00596	-1.349099
Ar	-526.8178	0.00609	-1.731965
K	-599.1648	0.00688	-2.196484
Fe	-1262.444	0.00948	-8.539770
Co	-1381.415	0.00995	-10.03890
Ni	-1506.872	0.01044	-11.72724
Cu	-1638.951	0.01092	-13.61988
Br	-2572.443	0.01329	-30.49500
Ag	-5197.519	0.01911	-107.6317
Hg	-18409.01	0.03450	-1021.865

[a] The total atomic energy in this approximation is given by
the sum of the tabulated three contributions. Note that it
does not involve correlation energy. All entries are in relative values E/E_h (see Appendix A). The configurations
correspond to a normal aufbau process and they coincide in
most cases with those proposed for the ground states from
experimental information.

initio model so defined, let us examine its inherent approximations. The adoption of the nonrelativistic Hamiltonian (1.1) brings about that the relativistic effects are not accounted for. Computationally, the estimation of relativistic effects in molecules is difficult so that almost all our knowledge is based on results for atoms (see, however, new developments noted in Section 5.J.). From Table 1.1 it is seen that except for several of the lightest atoms the relativistic energies are much larger than the energy changes observed in chemical reactions. Fortunately the largest contributions to the relativistic energy are due to inner electronic shells. Table 1.2 shows that the relativistic contribution is approximately constant regardless the number of valence shell electrons. It may therefore be as-

T a b l e 1.2

Relativistic energies (E/E_h) of selected atoms and their ions[2]

Atom	X^-	X	X^+	X^{2+}	X^{3+}	X^{4+}
C	-0.01345	-0.01383	-0.01408	-0.01409	-0.01344	-0.01247
F	-0.08034	-0.08208	-0.08285	-0.08271	-0.08524	-0.08606
Cl	-1.33799	-1.34910	-1.35005	-1.34928	-1.35351	-1.35422
Br		-30.49500	-30.49723	-30.49179	-30.50494	-30.50319
I		-179.6522	-179.6553	-179.6433	-179.6661	-179.6604

sumed that the total atomic relativistic energy is almost independent of the atomic electronic state and the chemical environment in molecules, which implies cancellation of relativistic effects in chemical processes. According to Kołos[4] the correction of the total energy of H_2^+ for relativistic effects amounts to -1.60 cm^{-1} whereas with the dissociation energy it is only 0.14 cm^{-1}. Data for H_2 are presented in Table 1.3. This table also gives us evidence on the validity of

T a b l e 1.3

Dissociation energy (D_o) of the H_2 molecule[4] (cm^{-1})

Born-Oppenheimer approximation	36112.2
Adiabatic approximation	36118.0
Relativistic correction to D_o	-0.5
Adiabatic approximation with corrections	36117.3
Experiment	36116.3-36118.3

the Born-Oppenheimer approximation. The case of the H_2 molecule in the ground state is representative. This molecule contains the lightest nuclei and therefore represents the case for which the approximation separating motions of nuclei from motions of electrons should be the least substantiated. Nevertheless the result in Table 1.3 is very satisfactory. One should keep in mind, however, that the nonadiabatic effects may be important in the cases, in which avoided curve crossing is involved or where two electronic states of different symmetry (and/or multiplicity) are close in energy. This is particularly topical for detailed spectroscopical studies and theoretical treatments of the dynamics of some chemical processes.

To conclude the analysis of the approximations noted, it is possible to state that, from the point of view of chemical applications, they represent very subtle effects. The experience shows that if ab initio calculations are in disagreement with experiment, it is in most cases not due to the approximations noted. As will be shown in next chapters, the crucial point in ab initio calculations is the basis set effect and, if the calculations are at the SCF level, also the correlation energy. The only exceptional case we shall meet in this book concerns ionization potentials for core electrons in molecules containing heavy atoms. Here the relativistic effects are very important.

2. Basis Set

2.A. Fundamental Concepts and General Description

The most widely used quantum chemical methods are based on the idea of the expansion of atomic or molecular orbitals as a linear combination of basis functions

$$\varphi_i = c_{i1}\chi_1 + c_{i2}\chi_2 + \cdots + c_{in}\chi_n \qquad (2.1)$$

where φ_i is a molecular or atomic orbital and χ are selected Slater--type, Gaussian or some other functions that are referred to as ba-sis set. If φ_i is a molecular orbital (MO), the expansion (2.1) is called MO-LCAO (linear combination of atomic orbitals), though with ab initio calculations it has a broader meaning than implied by the abreviation. The expansion coefficients $c_{i1}, c_{i2}, \ldots, c_{in}$ for $i = 1,\ldots, n$ (n is the number of basis set functions) are variational parameters and they are determined by the solution of the SCF problem (see Chapter 3). It is clear that in calculations on atoms the expansion (2.1) must contain at least one function for each occupied atomic orbital. Such a basis set is called minimum basis set. For example, for oxygen the minimum basis set is 1s, 2s, $2p_x$, $2p_y$ and $2p_z$.

The LCAO approximation (2.1) may be taken as the expansion of a function as a series. As for any method which contains such an expansion, it may be assumed that upon augmenting the expansion by some other functions one arrives at a superior description of atomic (or molecular) orbitals because the number of variational parameters is increased. Examine now a basis with a doubled number of functions, i.e. a basis set in which each atomic orbital is represented by two functions. The result of this basis set extension is presented in Table 2.1 for the atoms from the first to fourth rows of the periodic system. The results for the fifth-row atoms are presented in Table 2.2. The entries in the first columns of the two tables are total energies of atoms given by the SCF calculations with the minimal basis set of Slater-type orbitals (STO, see Section 2.B.). The entries in the second columns are energies given by the calculations in which each STO was replaced by two Slater-type functions with the exponents so optimized to give the minimum total energy. From Tables 2.1 and 2.2, it is seen that a twofold enlargement of the basis set gave

T a b l e 2.1

Total SCF energies[5] for atoms with Z = 2-36

Atom	State	Minimum STO basis set	DZ STO basis set	Hartree-Fock limit[a]
He	^1S	-2.8476563	-2.8616726	-2.8616799
Li	^2S	-7.4184820	-7.4327213	-7.4327256
Be	^1S	-14.556740	-14.572369	-14.573021
B	^2P	-24.498369	-24.527920	-24.529057
C	^3P	-37.622389	-37.686749	-37.688612
N	^4S	-54.268900	-54.397951	-54.400924
O	^3P	-74.540363	-74.804323	-74.809370
F	^2P	-98.942113	-99.401309	-99.409300
Ne	^1S	-127.81218	-128.53511	-128.54705
Na	^2S	-161.12392	-161.84999	-161.85891
Mg	^1S	-198.85779	-199.60701	-199.61461
Al	^2P	-241.15376	-241.87307	-241.87668
Si	^3P	-288.08996	-288.85116	-288.85431
P	^4S	-339.90988	-340.71595	-340.71869
S	^3P	-396.62762	-397.50229	-397.50485
Cl	^2P	-458.52369	-459.47960	-459.48187
Ar	^1S	-525.76525	-526.81511	-526.81739
K	^2S	-598.08987	-599.16241	-599.16446
Ca	^1S	-675.63390	-676.75594	-676.75802
Sc	^2D	-758.40414	-759.72637	-759.73552
Ti	^3F	-846.81561	-848.38875	-848.40575
V	^4F	-940.97197	-942.85728	-942.88420
Cr	^5D	-1041.0063	-1043.2709	-1043.3095
Mn	^6S	-1147.1067	-1149.8140	-1149.8657
Fe	^5D	-1259.0855	-1262.3715	-1262.4932
Co	^4F	-1377.3744	-1381.3205	-1381.4142
Ni	^3F	-1502.0487	-1506.7517	-1506.8705
Cu	^2S	-1632.3354	-1638.7496	-1638.9628
Zn	^1S	-1771.1509	-1777.6699	-1777.8477
Ga	^2P	-1916.5167	-1923.1110	-1923.2604
Ge	^3P	-2068.5139	-2075.2284	-2075.3591
As	^4S	-2227.2649	-2239.1207	-2234.2382
Se	^3P	-2392.7274	-2399.7563	-2399.8658
Br	^2P	-2565.1131	-2572.3415	-2572.4408
Kr	^1S	-2744.5197	-2751.9613	-2752.0546

[a] The estimated limits differ somewhat from those (Ref. 2) given in Table 1.1.

T a b l e 2.2

Total SCF energies[5] for atoms with Z = 37-54

Atom	State	Minimum STO basis set	DZ STO basis set
Rb	^2S	-2930.6931	-2938.2708
Sr	^1S	-3123.7176	-3131.4652
Y	^2D	-3324.7806	-3331.6538
Zr	^3F	-3531.3181	-3538.9632
Nb	^4F	-3475.4826	-3753.5211
Mo	^5D	-3967.0398	-3975.4131
Tc	^6S	-4196.0536	-4204.7591
Ru	^5D	-4432.3604	-4441.4569
Rh	^4F	-4676.2637	-4685.7699
Pd	^3F	-4927.8059	-4937.7504
Ag	^2D	-5187.0705	-5197.4836
Cd	^1S	-5454.1908	-5465.0971
In	^2P	-5729.0986	-5740.1392
Sn	^3P	-6011.6720	-6022.9057
Sb	^4S	-6302.0043	-6313.4618
Te	^3P	-6600.0387	-6611.7621
I	^2P	-6905.9462	-6917.9602
Xe	^1S	-7219.7923	-7232.1189

lower total energies in all cases. The basis set, in which two functions are assigned to each atomic orbital, is referred to as "double-zeta" basis set. In our case the basis set may be called DZ STO.
For Slater-type orbitals the exponents are usually denoted by ζ and
DZ is therefore to imply that to each atomic orbital two exponents,
i.e. two functions, are assigned. If atomic orbitals are represented
by more than two functions, the basis set is referred to as "extended" basis set. The same classification of basis sets is also used
for molecules. The basis set for a molecule is composed from the basis sets of pertinent atoms and according to the nature of these a-
tomic basis sets it is called minimum, double-zeta or extended. For
molecules, the extended basis sets usually also contain functions
which correspond to atomic orbitals with higher azimuthal quantum
numbers than those corresponding to atomic orbitals occupied in the
ground states of atoms. Such functions are called polarization functions. Of course, polarization functions do not affect the total SCF
energy of atomic ground states. For example, polarization functions

for hydrogen are of the p, (d, ...) types, for oxygen of d, (f, ...)
types, and for iron of f, (g, ...) types. Typically, polarization
functions are used in DZ+P (double zeta plus polarization) basis sets.
For hydrogen, the DZ+P basis set may be expressed as (2s1p) which
means two s-type functions, one p_x-type function, one p_y-type function
and one p_z-type function and for oxygen as (4s2p1d) which means four
s-type functions, two sets of p_x, p_y and p_z functions and one complete
set of d-functions (xy, xz, yz, $x^2 - y^2$ and $3z^2 - r^2$). In the usual
notation for molecules, the DZ+P basis set for H_2O is expressed as
(4s2p1d/2s1p), where the symbols standing before the slash refer to
oxygen and symbols after the slash to hydrogen. The choice of the term
"polarization functions" has not been very felicitous, but its use is
nowadays generally accepted. The term should reflect the fact that inclusion of polarization functions into the basis set permits a superi-
or description of the electron density distribution in molecules, i.e.
the "orbital polarization".

As the number of functions is increased, description of molecular
orbitals becomes better and better and the total SCF energy decreases.
In the limiting case of an infinite expansion (2.1), the set of functions $\{\chi\}$ is complete and the function φ_i is expressed "accurately"
through the expansion (2.1). Within the framework of the SCF approach,
the description of molecular orbitals cannot be further improved and
the energy cannot be lowered. This limiting value of energy is referred to as the Hartree-Fock limit. A typical dependence of the total SCF energy on the size of basis set is presented in Fig. 2.1 for
the water molecule using the STO basis sets. This dependence exhibits
some characteristic features which for ab initio molecular calculations may be generalized as follows:

i) Passage from the minimum to DZ basis set brings about a large
decrease in energy.

ii) Further basis set extension is effective only if polarization
functions are included.

iii) For larger than DZ+P basis sets, the energy already converges
slowly to the Hartree-Fock limit.

With ab initio calculations, the dependence of the cost of SCF
calculations on the size of the basis set is considerably more prohibitive than it is with semiempirical calculations. With semiempirical
methods, the evaluation of the necessary integrals over basis set
functions is very fast, so that the major portion of the computer
time consumed is due to the SCF procedure itself. Roughly speaking,
if n is the size of the basis set the computation time goes as n^2 to

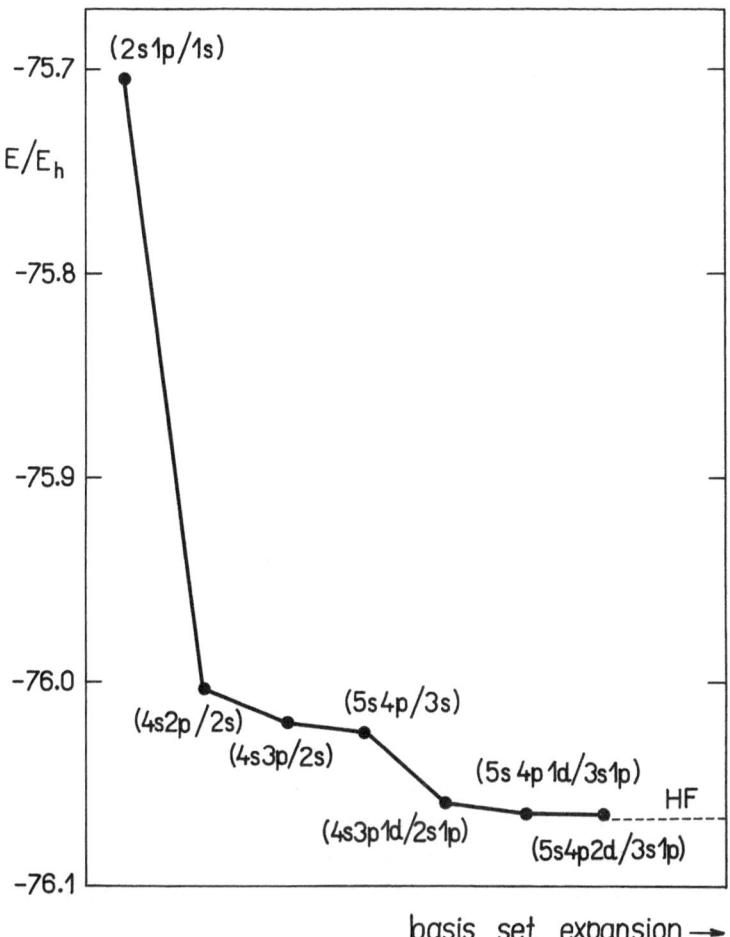

Figure 2.1
Dependence of the SCF energy of H_2O on the size of
the STO basis set[6-8]. The Hartree-Fock limit is indicated by the dashed line.

n^3, depending on the nature of the diagonalization procedure. The situation with ab initio calculations is different. For larger basis sets,
the most time-consuming step is the evaluation of two-electron integrals. If n is the number of basis set functions, then it is necessary
to compute generally $1/8(n^4 + 2n^3 + 3n^2 + 2n)$ nonequivalent two-electron integrals. As the basis set is enlarged, the computer time increases more rapidly by 1-2 orders in comparison to semiempirical methods. In Table 2.3 we present numbers of two-electron integrals that
are to be computed for the water molecule for several different basis

Table 2.3

Number of nonequivalent two-electron integrals for the
water molecule with different basis sets

Basis set	Number of functions (n)	Number of integrals $1/8(n^4+2n^3+3n^2+2n)$
(2s1p/1s)	7	406
(4s2p/2s)	14	5565
(4s3p/2s)	17	11781
(5s4p/3s)	23	38226
(4s3p1d/2s1p)	29[a]	94830
(5s4p1d/3s1p)	35[a]	198765
(5s4p2d/3s1p)	41[a]	371091

[a] For practical reasons it is usual to employ a set of
six d-functions (x^2, y^2, z^2, xy, xz, yz), which is
equivalent to five linearly independent d-functions
and a single very diffuse s-type function.

sets. From Fig. 2.1 and Table 2.3 two opposite trends may be noticed:
a slower energy convergence as a basis set becomes larger and a rapid
increase in the number of integrals as the number of basis set functions is increased. Obviously, some compromise must be made according
to the nature of the problem studied. It is evident that for small
molecules large basis sets may be used, whereas for larger molecules
only results for small basis sets are obtainable.

In contrast to semiempirical calculations, a typical feature of
ab initio studies is that a particular chemical problem is usually
treated by making use of several different basis sets. For molecular
calculations, basis sets of various size were suggested. Especially,
the choice of smaller basis sets (minimum, DZ) is rather abundant.
This situation may resemble the problem of empirical parameters in
semiempirical methods and the "nonempirical" nature of ab initio calculations may therefore appear questionable. The resemblance is, however, only apparent. The essential difference from semiempirical calculations is that for large basis sets no case is known of a dramatic
disagreement between theory and experiment (provided the electron
correlation is accounted for). In general, extensions of the basis
set lead to a superior description of all properties of the system.
Unlike in semiempirical methods, it is therefore possible to test directly the quality of the approach assumed by recalculating part of
calculations with a larger basis set (and by accounting for the correlation effects). An important feature of variational ab initio cal-

culations (SCF, CI) is the circumstance that the provided energy is an "upper bound" to the exact (experimental) energy.

Up to now we have assumed in this chapter the use of Slater-type orbitals. Actually, use may be made of any type of functions which form a complete set in Hilbert space. Since for practical reasons the expansion (2.1) must be always truncated, it is preferable to choose functions with a fast convergence. This requirement is probably best satisfied just for Slater-type functions. Nevertheless there is another aspect which must be taken into account. It is the rapidity with which we are able to evaluate the integrals over the basis set functions. This is particularly topical for many-center two-electron integrals. In this respect the use of the STO basis set is rather cumbersome. The only widely used alternative is a set of <u>Gaussian-type functions (GTF)</u>. The properties of Gaussian-type functions are just the opposite - integrals are computed simply and, in comparison to the STO basis set, rather rapidly, but the convergence is slow.

STO and GTF basis sets, which are the most important for practical calculations, will be treated separately in the next sections. For the sake of completeness we comment here on some other possibilities.

Some time ago the one-center expansions (OCE) seemed to be promising[9]. In this expansion a molecular orbital is expressed by functions which are all centered on one point, usually in the center of the molecule. The computation of one-center integrals is very fast, but the OCE approach requires the inclusion of functions with high azimuthal quantum numbers. The method is suited for symmetrical molecules of the AH_n type. However, use of OCE remained rather limited. Some other examples of less common basis set functions are elliptical functions[9-12] and Gaussian functions of special types such as cusped[13], even-tempered[14], Hermitian[15,16] Gaussian basis functions and those including interelectronic coordinates[17,18].

2.B. Slater-Type Orbitals

In the original meaning of the MO-LCAO approximation, the molecular orbitals are constructed from the combinations of orbitals of atoms constituting the molecule. For molecular orbitals so formed it may be assumed that their quality will reflect the quality of atomic orbitals used. But the AO description of atoms is accurate only for the hydrogen atom and the so-called hydrogen-like atoms, i.e., for systems with one electron and the nucleus with charge Z. For other atoms the AO description is approximate. For this reason the hydrogen-like functions may appear as a good candidate for basis sets suitable for calculations of molecules. They have the following form

$$\Psi_{n,\ell,m}(r,\vartheta,\varphi) = N_{n,\ell}\left(\frac{2Zr}{n}\right)^{\ell} L_{n+\ell}^{2\ell+1}\left(\frac{2Zr}{n}\right)e^{-Zr/n} Y_{\ell,m}(\vartheta,\varphi) \qquad (2.2)$$

where $N_{n,\ell}$ is a normalizing factor, $L_{n+\ell}^{2\ell+1}(x)$ are associated Laguerre polynomials of argument x, $Y_{\ell,m}(\vartheta,\varphi)$ are spherical harmonics, r, ϑ, φ are spherical coordinates of the electron relative to the nucleus and n, ℓ, m are quantum numbers, all integral, with n > 0, n - 1 $\geq$ ℓ $\geq$ 0, $\ell \geq m \geq -\ell$. Functions (2.2) form a complete set, they are normalized and mutually orthogonal. However, for actual calculations they are impractical because of their rather complicated form. Slater suggested functions of a similar form that are still approximate for many-electron atoms but that are considerably simpler. They are called Slater--type orbitals (STO) and have the following form[19,20]

$$\Phi_{\zeta,n_s,\ell,m}(r_A,\vartheta,\varphi) = \left[(2n_s)!\right]^{-1/2}(2\zeta)^{n_s+1/2} r_A^{n_s-1} e^{-\zeta r_A} \qquad (2.3)$$

Here n_s denotes "effective quantum number", exponent ζ is an arbitrary positive number, r_A, ϑ, φ are polar coordinates for a point with respect to the origin A in which the function (2.3) is centered. Apart from the first two terms that represent a normalizing factor, the function (2.3) is closely related to hydrogen-like orbitals. For the hydrogen 1s orbital the function $\Phi_{0,0,0}$ is identical with $\Psi_{0,0,0}$, if we assume ζ = Z/n. However, it should be recalled that in contrast to hydrogen-like orbitals STO's are not mutually orthogonal. Another essential difference is in the number of nodes. Hydrogen functions have (n - ℓ - 1) nodes, whereas STO's are nodeless in their radial part. Alternatively, the STO may be expressed by means of Cartesian coordinates as follows

$$\Phi_{\zeta,n_s,i,j,k} = N_{\zeta,n_s,i,j,k} r_A^{n_s-1-i-j-k} x_A^i y_A^j z_A^k e^{-\zeta r_A} \qquad (2.4)$$

where $x_A = (x - A_x)$, etc., A_x, A_y, A_z being Cartesian coordinates for

a point in which the function is centered. N is a normalizing factor. Note that any STO may be written as a linear combination of
functions of the type (2.4).

According to Slater[19], the exponents ζ are determined by means of
the formula

$$\zeta = \frac{Z - S}{n_s} \tag{2.5}$$

where (Z - S) is the effective nuclear charge and S is a constant to
account for the effect of screening part of the nuclear charge by the
other electrons. On the basis of a series of atomic calculations Slater derived empirical rules that permit n_s and ζ to be determined for
any orbital of any atom (see standard textbooks on quantum chemistry).
Functions with parameters determined according to Slater's rules are
called Slater orbitals and they are used in all current semiempirical
MO methods. With ab initio calculations another approach is preferred.
The n_s parameters are usually fixed at the values of principal quantum numbers and the exponents ζ are optimized to give minimum SCF a-
tomic energy. Optimum exponents given by the minimum basis set are
different from those provided by Slater's rules. Apart from the origin of exponents, any function of the type (2.3) is called STO. Here
no confusion can arise because with ab initio calculations the origin
of the exponents used is always specified.

As we have shown in Section 2.A., the extension of the minimum to
DZ basis set brings about the decrease in energy. For this extension,
of course, a new set of exponents is required. Exponent optimization
of STO basis sets, particularly extended basis sets, is not simple,
though all one-center integrals appearing in the SCF atomic problem
are expressible in closed analytical forms. It involves the optimization of a number of nonlinear parameters, in addition to the same
number of linear parameters in eqn. (2.1). Excellent mathematical a-
nalyses of the problem and the corresponding computer program were
communicated by Roothaan and Bagus[21]. The program was extended by
Clementi and Raimondi[22].

STO exponents ζ, optimum for atoms, are usually used without
change (not for hydrogen) for calculations of molecules. Usually no
use is made of the coefficients standing in the expansions of individual atomic orbitals over basis set STO's, i.e. usually no attempt
is made at what with Gaussian basis sets is called contraction (see
Section 2.D.). This is not profitable for STO basis sets because

their flexibility would be lost[23]. It should be realized that the coefficients $c_{i\mu}$ standing at individual basis set functions are different in AO expansions for atoms and MO expansions for molecules. The same holds of course also for exponents. Here, however, the optimization is too difficult. Since we are forced to use for molecular calculations exponents optimum for atoms, the following questions may be asked[23,24]. To what degree do atomic orbitals change when molecules are formed? What sequence, combination and extent of optimization of the orbital exponents ζ for molecules is necessary and useful? How many STO symmetry basis functions are needed to adequately represent each molecular orbital symmetry type? What kind of functions (1s, 2s, 3s, ... 2p, ... 3d, ...) is needed and what number of each type should be included in the basis set? Naturally, answers to these questions were looked for in calculations on diatomic molecules.

Diatomic molecules are still rather simple systems suitable for clarifying problems encountered in polyatomic molecules. Moreover, they give us additional information not obtainable from atomic calculations. Polarization functions represent a typical example. Since they correspond to orbitals unoccupied in atomic ground states, their exponents cannot be estimated from atomic calculations. Among the calculations on diatomic molecules, very useful information on the problem of basis set composition was contributed by calculations reported by Cade, Huo, Wahl, Sales, Liu, Yoshimine and others[24-28]. Consider for example the results for the nitrogen molecules. Fig. 2.2 presents the dependence of the total energy and its components (ki-

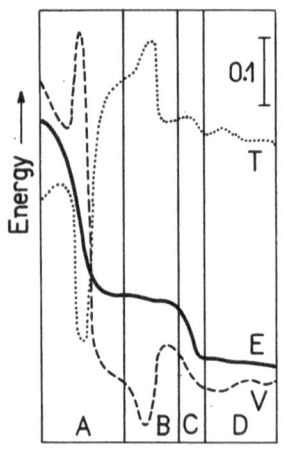

Basis set expansion →

Figure 2.2

Progression in the STO basis set[24] for the N_2 molecule and the total SCF (full line), potential (dashed line) and kinetic (dotted line) energies in arbitrary scales (the 0.1 (E/E_h) energy range indicated at the right top). A - minimum to DZ basis sets; B - extended sp basis sets; C - sets augmented with d-functions; D - sets augmented with d and f functions.

netic and potential energies) on the basis set quality. The basis set is extended step-wise starting from the best minimal molecular orbitals set, the STO exponents being optimized in N_2 in each step. As regards the total energy, the overall trends are qualitatively the same as those for the water molecule (see Fig. 2.1) and may be assumed to be also valid for polyatomic molecules. Rather disturbing in Fig. 2.2 is the erratic behavior of the kinetic and potential energy values. Evidently, the components of the total energy are more sensitive to the quality of the basis set than the total energy itself. This is sometimes used for testing the quality of basis sets, inasmuch as for Hartree-Fock wave functions the virial ratio -V/T becomes equal to two.

For molecules the optimization of exponents is very important for small basis sets but it can never alone absorb the deficiency due to a lack of expansion functions. The number and secondly the kind of STO basis set functions (i.e. s, p, d, ... type) are the most important considerations. The decreasing importance of the exponent optimization as the basis set grows is observable from Table 2.4.

T a b l e 2.4
Effect of the optimization of STO exponents on the SCF energy (E/E_h) of the HF molecule[a]

Basis set	Initial energy[b]	Optimized energy	Energy gain on optimization
(2s1p/1s)	-99.49143	-99.53614	0.04471
(3s2p/1s1p)	-99.97222	-100.03507	0.06285
(5s4p/3s1p)	-100.05587	-100.06051	0.00464
(5s4p2d1f/3s1p1d)	-100.06939[c]	-100.07030	0.00091

[a] Ref. 26; exponents of p, d, f functions contributing to σ and π orbitals are optimized independently (anisotropic optimization).
[b] Energy given by the exponents optimum for atoms.
[c] Estimated exponents for polarization functions.

Another important result of the papers by Cade, Huo, Wahl and collaborators is the finding that it is not necessary to reoptimize STO exponents for different internuclear distances. If we are not far from the distance at which the exponents were optimized, the energy gain upon exponent reoptimization is very small.

It certainly need not be emphasized that the exponent optimiza-

tion for molecules is a very time-consuming process. For large basis
sets the energy gain is small so that it is sufficient to estimate
or to optimize exponents of polarization functions and to employ exponents from atomic calculations for the other functions. With small
basis sets, a basis set extension is more economic than the exponent
optimization. Thus the question may asked whether the exponent optimization is warranted at all. A plausible answer is that it is in
certain cases. We note here on such a case of its possible use in
treatments of large molecules. It should be realized that in ab initio
applications to large molecules one is forced to make use of small,
say, minimum or DZ basis sets. A standard procedure would be to take
the optimum exponents for atoms. Alternatively, one may perform the
exponent optimization for a small molecule and then to employ the exponents so optimized for a larger molecule with a similar bonding environment. The idea used here is the assumed transferability of optimized exponents for atoms in chemically similar molecular environments. The basis set optimized in this way permits us to take into
account the nature of the perturbing environment in which the atom
finds itself in a molecule. In Table 2.5 we present an example of
such a treatment[29] for C_2H_6 and H_2O_2.

Table 2.5
Use of minimum STO basis sets with exponents optimized
in small molecules[29]

Exponents	Total energy (E/E_h)	
	C_2H_6 (staggered)	trans-H_2O_2
Slater exponents	-78.9912	-150.1467
Optimized for CH_4	-79.0980	-
Optimized for H_2O	-	-150.2232
Fully optimized	-79.0999	-150.2353

With barriers to internal rotation in C_2H_6 and H_2O_2, the exponent
optimization noted in Table 2.5 brought about no improvement. Actually, use of optimum exponents for H_2O instead of Slater exponents resulted in a worse agreement with experimental results for H_2O_2. Difficulties in accounting for the barrier to internal rotation in H_2O_2
are not compatible with the problems of minimum basis set optimization
and they will be discussed later. We note it here just to point out

that the energy lowering upon exponent optimization should not be overemphasized.

For STO calculations reported in the literature a variety of basis sets was used. Optimum exponents for minimum and DZ atomic basis sets were reported by Clementi and coworkers[5,22]. Among larger basis sets the "nominal basis sets" of Bagus et al.[30], carefully optimized for He-Kr, are widely used. A nominal basis set consists of two basis functions per shell, except that three basis functions are used for the first p-shell and four for the d-shell. For the transition-metal atoms and their dipositive ions, Huzinaga et al. developed[32] (10s6p5d) and (7s6p5d) basis sets, respectively.

2.C. Gaussian-Type Functions

Gaussian functions are known and used in various branches of physics and chemistry. Their use for quantum chemical calculations was first suggested by Boys[31] in the following form

$$g(\alpha,A,i,j,k) = N_{\alpha,i,j,k} \, x_A^i y_A^j z_A^k \, e^{-\alpha r_A^2} \tag{2.6}$$

Here α is positive parameter (orbital exponent), i, j, k are integers ≥ 0. Functions with i = j = k = 0 are referred to as s-type functions, functions with i = 1, j = k = 0 as p_x-type functions, etc. Note that functions (2.6) do not possess principal quantum numbers. $N_{\alpha,i,j,k}$ is a normalizing factor

$$N_{\alpha,i,j,k} = \left[\left(\frac{\pi}{2\alpha}\right)^{3/2} \frac{(2i-1)!!(2j-1)!!(2k-1)!!}{2^{2(i+j+k)} \, \alpha^{(i+j+k)}} \right]^{-1/2} \tag{2.7}$$

The functions (2.6) are centered in the point A with the coordinates A_x, A_y, A_z and r_A denotes the distance between a point A and a variable point with coordinates x, y, z:

$$r_A^2 = (x - A_x)^2 + (y - A_y)^2 + (z - A_z)^2 \tag{2.8}$$

$$r_A^2 = x_A^2 + y_A^2 + z_A^2 \tag{2.9}$$

It is customary to call the functions (2.6) Gaussian-type functions (GTF). In the form (2.6) the angular dependence is expressed by powers of x, y, z coordinates and the functions are accordingly referred to as Cartesian GTF's. Another currently used expression is given by

$$g(\zeta_g, A, n_g, \ell, m) = N \; r_A^{n_g-1} \; e^{-\zeta_g r_A^2} \; Y_{\ell, m}(\vartheta, \varphi) \tag{2.10}$$

where the normalizing factor is

$$N = \left[\frac{2^{2n_g + 3/2}}{(2n_g - 1)!!\sqrt{\pi}} \right]^{1/2} \zeta_g^{(2n_g+1)/4} \tag{2.11}$$

and n_g and ζ_g are analogues of the principal quantum number and orbital exponent of STO's.

In principle, since GTF's form a complete set, the exact molecular orbitals may be expressed in terms of them. Unfortunately the behavior of the Gaussians near the nucleus and far away is incorrect, so many more GTF's than STO's are needed to approximate the exact orbitals to the same degree of accuracy. This point will be discussed in the next sections. Here we note only once again that this disadvantage is overweighed by the ease with which the integrals over GTF's are computed.

As regards the use of GTF's as a basis set for molecular calculations we have several possibilities.

(i) As with the STO basis set, atomic orbitals are expanded in the series over GTF's and their exponents optimized to give the lowest SCF atomic energy. GTF's with optimized exponents may be used directly in calculations on molecules or, less typically, their exponents may be reoptimized in molecules. No use is made of expansion coefficients given by atomic calculations. Such a procedure is referred to as the calculation with the uncontracted GTF basis set. At the present time, the molecular calculations with uncontracted basis sets are very rare.

(ii) In contrast to (i), part of the expansion coefficients given by the atomic SCF calculation is used in molecular calculations. These coefficients are not subjected to variation in the SCF calculation. Instead we consider a fixed grouping of several pertinent GTF's as one basis function. Such a function is called contracted and it is denoted by CGTF. By making use of CGTF's the number of variational parameters

of the SCF procedure is decreased. This compensates to a certain degree the requirement for a large number of GTF's in the basis set.
Compared to (i) the flexibility of the basis set is of course lower
but, as will be shown in Section 2.D., for justifiable contractions
the loss in flexibility is rather small.

(iii) The STO's are expanded into the series of GTF's by means of
a least squares fit. The SCF calculations of molecules (or atoms) are
then performed with this STO basis set, in which each STO is simulated by a fixed combination of several GTF's. This approach is very
important and it will be treated separately in Section 2.E.

(iv) The expansion over GTF's is made for atomic SCF orbitals.
The molecular calculations are then made with the so expressed AO basis set. This approach is of no practical significance because it
gives us no advantage over the approach (iii).

This section is devoted to the point (i). Early calculations in
which each atomic orbital was represented by a single Gaussian were
rather discouraging[33,34]. This is comprehensible because exponentials,
not Gaussians, form natural solutions to the central-field problem.
Next calculations[35-37] in which expansion over several GTF's was made
per atomic orbital established utility of the idea of using Gaussians
as basis set functions. The energy convergence to the Hartree-Fock
limit is presented in Table 2.6 for the hydrogen atom and the hydrogen molecule. It is seen that the energy convergence is rather satis-

T a b l e 2.6

Dependence of SCF energy on the number of GTF's for
the hydrogen atom and the hydrogen molecule

GTF's per H atom	Basis set[a]	SCF energy $(-E/E_h)$	
		H	H_2[b]
2	2s	0.485813	1.09431
4	4s	0.499277	1.12655
6	6s	0.499940	1.12834
8	8s	0.499991	1.12851
6	5s1p	–	1.13269
7	5s2p	–	1.13346
Hartree-Fock limit[c]		0.5	1.13364

[a] The s sets from Ref. 38.
[b] Refs. 39,40.
[c] Ref. 41.

factory. For example, even for two GTF's per hydrogen atom the energy of H_2 is lower than the energy[42] $E/E_h = -1.09092$ given by the minimum STO basis set.

In the early stage of calculations with GTF basis sets, the exponents were determined by an approximate optimization[43,44]. They were constrained to be in geometric progression[+] so that the ratio between adjacent exponents was constant. In the optimization procedure two parameters, the exponent of the "center Gaussian" and the ratio of adjacent exponents, were varied. In this way the problem of many-parameter optimization was avoided. This procedure was used systematically by Csizmadia and coworkers[45] who tabulated GTF basis sets for atoms Li-Ne. They also reviewed calculations with uncontracted GTF basis sets. The experience acquired up to that time has shown that basis sets between two and three times the size of a STO basis set would give comparable results.

Approximate optimization of basis sets was very soon replaced by accurate procedures. A pioneering work in this field was made by Huzinaga and coworkers[38,46]. Huzinaga[38] investigated the usefulness of GTF's as basis set functions for large-scale molecular calculations in two different directions. The first of them aimed to obtain approximate expansion of STO's in terms of GTF's. This approach belongs to the category (iii) of our classification given above and it will be discussed in Section 2.E. The second approach is based on SCF atomic calculations with GTF's and belongs therefore to the category (i) of our classification. Huzinaga made use of the optimization program of Roothaan and Bagus[21] which was modified to accommodate Gaussian-type basis functions instead of Slater-type basis functions. He obtained accurate basis sets of the sizes (9s5p) and (10s6p) for atoms Li-Ne. Together with the hydrogen basis sets, obtained by a fit of the 1s STO by 2-10 GTF's, they represent an important and widely used source of basis sets up to the present time.

Nowadays optimized GTF basis sets of various size are available for most atoms of the periodic system. Since they represent a starting point for developing contracted basis sets, they will be noted in more detail in the following section for the sake of compactness.

[+] A revival of the idea of choosing exponents by means of geometric progression, though in a somewhat different form, is met with even--tempered Gaussian basis sets[14].

2.D. Contracted Gaussian Basis Sets

If a GTF basis set is to be used with the aim of giving better results than a minimum STO basis set for molecules composed from first-
-row atoms and hydrogen, a possible choice might be a (7s3p/3s) basis
set. For larger molecules, however, its size rapidly becomes formid-
able. For example, for benzene it would contain 114 functions and the
SCF calculation would be rather cumbersome due to a high number of
respective integrals. Moreover, in spite of a considerable effort the
result would be hardly better than that, say, given by a DZ-STO basis
set. In this section it will be shown how this handicap inherent to
GTF basis sets may be reduced by means of the basis set contraction.
This technique was suggested almost simultaneously by several authors
[47-50]. Its essence will be demonstrated here by means of the example
of NH_3 calculation with the (7s3p/3s) basis set. The exponents assumed
for nitrogen are those optimized by Whitman and Hornback[51]. Their
s-set exponents and the expansion coefficients for 1s and 2s atomic
orbitals are listed in Table 2.7. The exponents assumed for hydrogen

T a b l e 2.7
Exponents and expansion coefficients of the
(7s) basis set for nitrogen[51]

	Expansion coefficients	
Exponents	1s	2s
1619.0	0.0059	-0.0013
248.7	0.0424	-0.0096
57.75	0.1820	-0.0422
16.36	0.4570	-0.1326
5.081	0.4412	-0.1897
0.7797	0.0342	0.5077
0.2350	-0.0089	0.6151

originate from the optimization by Huzinaga[38]. From the inspection of
expansion coefficients in Table 2.7 it follows that the first five s-
-type GTF's with the highest exponents contribute predominantly to
the 1s atomic orbital whereas the 2s atomic orbital is formed almost
exclusively by only the two last GTF's. Hence it may be assumed that
the quality of the atomic orbitals will be affected very little if we
impose a restriction on the five GTF's with the largest exponents

that the ratio of their coefficients should be the same in the 2s as in 1s orbital. This means that we treat these five functions as a grouping with fixed coefficients. Similarly, the two remaining GTF's with small exponents may be combined into a second group with the coefficients from the 2s orbital. In this way the (7s) basis set reduces to a minimum basis set in the following form

$$\chi_1 = 0.0059\ g_1 + 0.0424\ g_2 + 0.1820\ g_3 + 0.4570\ g_4 + 0.4412\ g_5$$

$$\chi_2 = 0.5077\ g_6 + 0.6151\ g_7 \tag{2.12}$$

where χ_1 stands for the 1s atomic orbital and χ_2 for the 2s atomic orbital. The functions χ_1 and χ_2 are called <u>contracted Gaussian-type functions (CGTF)</u>, whereas the original functions, g, are referred to as <u>primitive</u> Gaussians. Molecular SCF calculations are then performed with the CGTF basis set. The contraction (2.12) may be denoted as (12345)(67). We may make also contractions for the 2p nitrogen and 1s hydrogen orbitals, respectively, by grouping three p-type GTF's for nitrogen and three s-type GTF's for hydrogen. Upon doing contractions in this way, the number of basis set functions is the same as in the minimum STO basis set. In the usual notation a CGTF basis set is given in brackets so that our basis set would be specified most typically as a (7s3p/3s) primitive set contracted to [2s1p/1s]. Obviously, the (7s3p/3s) basis set may also be contracted in other ways. For example, one can arrive at the DZ basis set by the (1234)(5)(6)(7) contraction of the nitrogen s-set, the (12)(3) contraction of the nitrogen p-set, and the (12)(3) contraction of the hydrogen s-set. In an equivalent way this contraction may be expressed as (4,1,1,1; 2,1/2,1) where the numbers of primitive GTF's entering individual CGTF's are indicated in the order of their descending exponents. The effect of the contraction is presented in Table 2.8. The total energy and one-electron properties given by the DZ basis set are seen to compare well with those given by the uncontracted set. But with the minimal basis set, considerable differences are obtained for some properties.

Generally, we define a contracted basis function by

$$\chi_i = N_i \sum_j c_{ji} g_j \tag{2.13}$$

where $\{g_j\}$ denotes the normalized primitive set and $\{\chi_i\}$ the contracted set. N_i is a normalizing factor of the contracted function χ_i.

T a b l e 2.8

Energy and some one-electron properties[a] of NH_3 calculated with (7s3p/3s) basis set and its contractions to the DZ and minimum basis sets

Property		(7s3p/3s)	[4s2p/2s]	[2s1p/1s]
Total energy E_{SCF}		−56.098387	−56.090590	−55.950048
Potential $\langle 1/r_N \rangle$ el.		−19.9362	−19.9314	−20.0231
	tot.	−18.3669	−18.3621	−18.4538
$\langle 1/r_H \rangle$ el.		−5.3637	−5.3578	−5.1854
	tot.	−1.0500	−1.0442	−0.8717
Electric field E_z (N)		0.1440	0.1985	0.2387
E_x (H)		0.1112	0.1162	0.1724
E_z (H)		−0.0189	−0.0183	−0.0342
Dipole moment μ		−0.8221	−0.8269	−0.7926
Quadrupole moment[b] Θ_{xx}		0.8499	0.8206	0.6488
Θ_{zz}		−1.6998	−1.6412	−1.2976

[a] Calculated at the experimental geometry: r_{NH} 1.9117; $\measuredangle$ HNH 106.7°. All entries are in relative values (see Appendix A). Exponents and contraction coefficients for nitrogen and hydrogen according to Ref. 51 and Ref. 38, respectively.
[b] Relative to the center of mass.

Contracted functions are constructed from primitives that are of the same symmetry type (i.e. s-type GTF´s, p-type GTF´s, etc.) and that are centered on the same nucleus. Contractions of a more general type are conceivable but they are not used.

In the case with NH_3, the contraction of the (7s3p/3s) primitive set to the minimum basis set was unequivocal. Unfortunately such a situation occurs rather rarely. Typically, there is some arbitrariness in the choice of the linear combinations. This holds particularly for larger basis sets where some primitives contribute significantly to both 1s and 2s orbitals of the first-row atoms. For second-row and further atoms the situation with s-type functions is even more complex. Ambiguity is also encountered with p-sets.

In testing a particular contraction for molecular calculations, the first natural step is the atomic SCF calculation[48,52,53]. However, it should be kept in mind that although a good contracted basis set must of necessity yield a satisfactory atomic energy, this alone is not sufficient. The basis set must also be flexible enough to allow for the changes in the valence atomic orbitals which occur upon molecular formation[53].

For p-set contractions of first-row atoms, the atomic calculations are completely useless. Since all p-type primitives of any
first-row atom contribute to a single set of atomic p-orbitals, the
atomic energy is independent of contraction and it therefore cannot
provide even a qualitative guidance for the contraction. The same
holds for hydrogen s-type functions. For this reason it is necessary
to test contractions of these GTF's directly in molecular calculations. Such tests were already made in early calculations with GTF
basis sets by several authors[53-58]. These papers did bring into view
a powerful means of increasing the efficiency of calculations with
Gaussian basis functions. Though contractions assumed in these papers
were more or less tentative, the results of their testing were very
important for a later formulation of general principles of basis set
contractions. From the experience acquired in these papers and from
a series of the own calculations on H_2O and N_2 with various contractions of the (9s5p/4s) basis set, Dunning[59] inferred the following
rules:

A. Those members of each group of basis functions, e.g., s, p, d,
etc., which are concentrated in the valence regions must be uncontracted.

B. If, within a particular group, one (or more) of the primitives
makes a substantial contribution to two (or more) atomic orbitals
with significantly different weights relative to the other functions
in the group, then this function must be left uncontracted.

From the first rule it follows that diffuse functions, i.e. with
low exponents, should be uncontracted because they are just the functions most concentrated in the interatomic regions of the molecule.
In contrast, the innermost primitives with largest exponents describe
a region which upon molecule formation remains largely atomic in character. For this reason these primitives are grouped into a single contracted function. The rather high number of primitives required for
the representation of inner shell functions is due to a poor behavior
of GTF's near the nucleus. To illustrate the utility of the second
rule, consider the following example. Let g_i be a primitive s-type
function contributing significantly to both 1s and 2s atomic orbitals.
Our task is to decide whether g_i may be grouped with two other primitives g_j and g_k. If the ratios of expansion coefficients in the 1s a-
tomic orbital, c_i^{1s}/c_j^{1s} and c_i^{1s}/c_k^{1s}, are markedly different from those
in the 2s atomic orbital, c_i^{2s}/c_j^{2s} and c_i^{2s}/c_k^{2s}, the function g_i should
be left uncontracted[60,61].

Up to now we have assumed that each primitive function is used in the contraction only once, i.e. each primitive function is only involved in one particular contracted function. The contractions of this type are referred to as segmented. In some cases it may occur that for a given primitive set the Dunning's rules suggest a higher number of CGTF's than we intended for a molecular calculation. Consider for example the (9s) and (10s) basis sets of Huzinaga[38] for the first-row atoms. The former contracts satisfactorily to a [4s] set, whereas the latter requires at least 5-6 functions[59,62]. A problem of avoiding an extension of the basis set without any loss in its quality may be bypassed in such a way that the GTF's which contribute significantly to several atomic orbitals are involved in the basis set several times and are used for each of these atomic orbitals[60,62]. Raffenetti[63] generalized this idea and he introduced the term general contraction for the scheme according to which all primitive functions may contribute to each of several contracted basis functions. If general contractions are applied to standard integral programs, it is mostly impossible to avoid a repeated calculation of integrals for primitives that occur in more than one contracted function. Raffenetti succeeded to develop a program[63] which is claimed to permit a general contraction at no expense in the integral computation time. Actually, his approach eliminates the problems encountered in contractions of atomic basis sets for use in molecular calculations.

To conclude this section in a practical way, we present a survey of Gaussian basis sets most widely used in actual calculations. The (9s5p) and (10s6p) basis sets of Huzinaga[38] for the first-row atoms were already noted. In later papers of Huzinaga and coworkers other basis sets were reported: (11s6p)[64], (11s7p)[65], (13s7p)[65] and of some other sizes[66]. Another important source of basis sets for first-row atoms is the collection by Duijneveldt[67] which covers basis sets from (4s2p) to (13s8p). Basis sets of various size were also developed by Whitman and Hornback[51]. Uniform quality basis sets (4s2p), (6s3p), (8s4p) and (10s5p) were suggested by Mezey et al.[68]. Among other basis sets used for the first-row atoms we note on the (7s3p) basis sets of Roos and Siegbahn[69] and Clementi[70]. As regards CGTF sets, very popular are the Dunning's contractions[59] of Huzinaga's (9s5p/4s) sets.

For the second-row elements, basis sets of various sizes were developed by Huzinaga and coworkers: (14s7p)[71] and (17s8p)[65] for Na and Mg and (14s10p)[71] and (17s12p)[65] for the other second-row atoms. For a more complete survey of their second-row basis sets, see Huzinaga's compilation[66]. Very useful basis sets of the size (12s9p) were devel-

oped by Veillard[72], who also tested their various contractions to
[6s4p]. It should however be noted that his contractions violate the
Dunning's rules[59]. Improved contractions of Veillard's sets to [6s4p]
were suggested later for Si, P, S[73] and Cl[60,73]. To complete a list
of basis sets for the second-row atoms the (10s6p) basis set of Roos
and Siegbahn[69] should be noted.

For the first-row transition-metal atoms Basch and coworkers[74] reported the (15s8p5d) basis set and its various contractions to the
minimum basis set [4s2p1d]. For the third-row atoms up to Zn the
(14s9p5d) basis sets are available[75] that are compatible with the
Veillard's (12s9p) sets for the second-row atoms. Their contractions
to the double-zeta basis set [8s4p2d] were also reported[75]. Two
smaller basis sets for the third-row atoms were reported by Roos and
coworkers[76]. The first of them, (9s5p3d), is so constructed as to be
compatible with a (7s3p) set for the first row atoms[51,69,70] whereas
the second, (12s6p4d), should be used in conjuction with a (8s4p) set
for the first-row atoms[51] and (10s6p) set for the second-row atoms[69].
Recently Gaussian basis sets for heavier atoms were reported[77,78].

2.E. Gaussian Expansion of Slater-Type Orbitals

As we have learned in Sections 2.B. and 2.C., STO's are well suited
for the use as basis set functions, but the evaluation of two-electron
integrals over them is time consuming. The opposite is true for GTF's.
The computation of integrals is relatively fast, but considerably
larger number of basis set functions is required. In this section we
discuss an approach which attempts to combine the merits of the two
types of basis sets. It preserves the nature of the STO basis set but
each STO is replaced by a linear combination of a small number of
GTF's, so that the integrals are evaluated actually over Gaussians.
Such a possibility was already discussed by several authors[47,79,80]
in 1960. In 1965 Huzinaga[38] used the variation procedure of McWeeny
for the Gaussian expansion of 1s, 2s, ... up to 3p STO's. In this approximate expansion, a STO centered in the point A (eqn.(2.3)), is
expressed in terms of K spherical GTF's (eqn.(2.10)) as follows

$$\Phi_{\zeta_s, n_s, \ell, m}(r_A, \vartheta, \varphi) = \sum_{i=1}^{K} c_i g(\zeta_{g_i}, A, n_g, \ell, m) \tag{2.14}$$

For all members of the expansion, a single set of A, n_g, ℓ and m values is used. It is sufficient to perform the optimization for $\zeta_s = 1$ in which case we obtain a set of optimum Gaussian exponents γ_i. Since the so-called scaling relation holds

$$\zeta_{g_i} = \zeta_s^2 \, \gamma_i \tag{2.15}$$

it is possible to arrive from the γ_i set, which is independent of ζ_s, at exponents of GTF's for any exponent ζ_s. The transformation does not refer to coefficients c_i.

As shown by Huzinaga, all s-type STO's are best expanded in terms of 1s-GTF's ($n_g = 1$), p-type STO's are best expanded in terms of 2p GTF's ($n_g = 2$). The adequacy of using GTF's with lower n_g values is very helpful to reduce complications in molecular integral calculations.

Alternatively, instead of the variational procedure a least squares fit of the Gaussian expansion to the STO may be used[46,81]. The two approaches were compared by Klessinger[82]. The method of least squares fitting became very popular owing to papers coming from Pople's group [83-86]. In the notation of the cited papers, STO's for first-row atoms have the form

$$\Phi_{1s}(\zeta_1, r) = (\zeta_1^3/\pi)^{1/2} \exp(-\zeta_1 r)$$

$$\Phi_{2s}(\zeta_2, r) = (\zeta_2^5/3\pi)^{1/2} \, r \, \exp(-\zeta_2 r)$$

$$\Phi_{2p}(\zeta_2, r) = (\zeta_2^5/\pi)^{1/2} \, r \, \exp(-\zeta_2 r) \cos \vartheta \tag{2.16}$$

The STO's are approximated by a linear combination of N 1s and 2p Gaussians:

$$\Phi_\mu'(\zeta, r) = \zeta^{3/2} \, \Phi_\mu'(1, \zeta r) \tag{2.17}$$

$$\Phi_{1s}'(1, r) = \sum_k^N d_{1s,k} g_{1s}(\gamma_{1s,k}, r)$$

$$\Phi_{2s}'(1, r) = \sum_k^N d_{2s,k} g_{1s}(\gamma_{2s,k}, r)$$

$$\Phi'_{2p}(1,r) = \sum_{k}^{N} d_{2p,k} g_{2p}(\gamma_{2s,k}, r) \tag{2.18}$$

Here g_{1s} and g_{2p} are the normalized primitive Gaussians

$$g_{1s}(\gamma, r) = (2\gamma/\pi)^{3/4} \exp(-\gamma r^2)$$

$$g_{2p}(\gamma, r) = (128\gamma^5/\pi^3)^{1/4} \; r \; \exp(-\gamma r^2) \; \cos\vartheta \tag{2.19}$$

Note that the same number of Gaussians, N, is used for each STO and that common Gaussian exponents $\gamma_{n,k}$ are shared between ns and np orbitals. Minimum basis sets of this type are referred to as STO-NG. We restricted ourselves here to the derivation of basis sets for the first-row atoms because the STO-NG basis sets for the second and higher row atoms may be obtained[85,86] along the same lines.

It should be noted that calculations may of course be made also with Gaussian expansions of STO's which are free of the restrictions imposed on the STO-NG basis sets. Nevertheless almost all our experience with Gaussian expansion originates from the results of calculations with STO-NG and related basis sets. So it seems to us tolerable to restrict the discussion just to the functions of these types.

Exponents γ and coefficients d in functions (2.18) were determined[86] by a least square fit for $\zeta = 1$. As in the treatment by Huzinaga, the passage to any ζ is made by means of the relation (2.15). The factor $\zeta^{3/2}$ stands in eqn. (2.17) for the renormalization of Gaussians after the formal change $r \rightarrow \zeta_s r$.

As N is increased, the convergence of STO-NG basis sets towards the STO results is rather satisfactory. A typical example is shown in Table 2.9. It is seen that the energy of atomization converges much

T a b l e 2.9
Convergence of STO-NG basis sets towards full STO results for CO[84]

Property	STO-3G	STO-4G	STO-5G	STO-6G	STO
Total energy	-111.2297	-112.0337	-112.2443	-112.3086	-112.3436
Energy of atomization	0.1999	0.1930	0.1924	0.1921	0.1921
Dipole moment	2.69	2.46	2.44	2.44	2.43

faster than the total energy. This indicates that the error introduced when the STO's are replaced by small Gaussian expansions is very similar in the molecule and in the separate atoms. As already noted in preceding sections, the error is largely associated with the inadequate behavior of GTF's near the nuclei, which results in a poor representation of inner shell orbitals.

From the results it follows that STO's expanded over GTF's may be treated as genuine STO's. As regards the selection of their exponents, we have several possibilities: to adopt the exponents ζ according to Slater rules, to use exponents optimum for free atoms, or to derive a set of exponents by a direct optimization in molecules. With STO-NG basis sets the last approach was favored. The exponents for inner shells (K shell for first-row atoms, K and L shells for second-row atoms) were fixed at values corresponding to free atom values. But for valence shell orbitals, the exponents were optimized for various STO-NG expansions in a series of molecules. As might be expected for a minimum basis set, the optimum exponents were rather different from those for free atoms. Optimum ζ values for a particular type of atom were found to be almost independent of the size of the Gaussian representation. Although the differences on going from one molecule to another were considerably larger, a single set of standard exponents was suggested[85] which should be appropriate for an average molecular environment. The standard exponents preserve the shell structure of the basis set, i.e. $\zeta_{2s} = \zeta_{2p}$ and $\zeta_{3s} = \zeta_{3p}$. This constraint reduces considerably the flexibility of the basis set, but it leads to significant improvement in timing of the integral evaluation (see Chapter 3). In spite of the restrictive features of STO-NG basis sets, the actual calculations proved their utility. Especially, predicted molecular geometries are reasonable, which together with the economy of calculations resulted in a widespread use of STO-NG basis sets, the most popular among them being STO-3G.

For different reasons, STO-NG basis sets give total energies inferior to those given by other Gaussian basis sets with approximately the same number of primitive Gaussians, though some other molecular properties are reproduced well (see for example the first two rows in Table 2.10). Two among possible reasons were already noted. The first is a poor representation of the 1s orbital. As shown by Klessinger[82], it is preferable to use more Gaussians for the 1s orbital than for the 2s orbital. The second is the restriction in the exponents of 2s and 2p orbitals. The main reason may however be assigned to the fact that STO-NG basis sets provide results converging in the limit as $N \to \infty$

T a b l e 2.10

Energies and dipole moments of H_2O given by different small basis sets

Basis set	Energy (E/E_h)	Dipole moment[a]
STO-3G[b]	-74.9659	5.63
STO-4G[c]	-75.5001	6.07
[2s1p/1s] [d,e]	-75.7317	7.34
[2s1p/1s] [d,f]	-75.8043	8.44
(2s1p/1s) [g]	-75.6568	4.85
4-31G[h]	-75.9084	8.41
[3s2p/2s] [i]	-76.0080	9.07
DZ-STO[j]	-76.0053	-
DZ-CGTF[i]	-76.0093	8.95

a In 10^{-30} Cm, experiment[87] 6.19.
b Refs. 84, 88.
c Ref. 89.
d Contracted[90] from (7s3p/3s).
e Exponents optimized for atoms.
f Exponents and contraction coefficients optimized for the molecule.
g Minimum STO basis[92], with Slater rules exponents.
h Ref. 91.
i Dunning's contraction[59] of the (9s5p/4s) set.
j Ref. 7.

to values corresponding to a pure minimum STO basis set. As regards the total energy, in many cases a minimum STO basis is poorer[54] than some minimum CGTF basis sets. For the water molecule, no better energy than -75.7055 can be obtained with the optimized STO minimum basis set[92] (see Table 2.10 in which a comparison is made for various STO and CGTF basis sets through the DZ size).

Obviously, if better flexibility is to be achieved, some decontraction of the valence shell functions must be made. In the popular 4-31G basis set[91] this was achieved by splitting the valence shell orbitals in the STO-4G set into two parts, the most diffuse primitive GTF being left uncontracted. In the 4-31G basis set, each inner-shell orbital is represented by a single function containing four GTF's and instead of valence-shell n_s and n_p orbitals we have $n_{s'}$ and $n_{p'}$ functions consisting of three Gaussians and $n_{s''}$ and $n_{p''}$ functions consisting of one Gaussian. Thus, 4-31G may be called a valence double--zeta basis set. Unlike with the STO-NG sets, the 4-31G exponents and contraction coefficients were obtained[91,93,94] by the minimization of atomic ground state energies. Again, the restriction $\gamma_{ns'} = \gamma_{np'}$ and $\gamma_{ns''} = \gamma_{np''}$ is assumed and a standard set of valence-shell scale factors, ζ' and ζ'', is suggested for use in molecular calculations.

The next attempts in the Pople's group for improvements in the basis sets were oriented to a better description of the inner-shell 1s orbitals of the first-row atoms[95]. The developed basis sets, 5-31G and 6-31G respectively, differ from 4-31G only by the representation of the inner shell which is now taken as a contraction of five and six Gaussians. With respect to the 4-31G basis set, improvement of inner-shell description leads[95] "to substantial lowering of calculated atomic and molecular total energies, but does not appear to alter calculated relative energies or equilibrium geometries significantly". This conclusion gives the justification for the use of the 4-31G basis set in chemical applications.

2.F. Polarization Functions

As regards the polarization functions the following two questions may be asked: (i) in which case the use of polarization functions is unavoidable; (ii) how to determine the optimum exponents of polarization functions for molecules.

The answer to the first question depends on several factors such as the accuracy which is to be achieved, the nature of the problem studied and the theoretical approach adopted. For example, from Figs. 2.1 and 2.2 we know that the energies cannot approach Hartree-Fock limits without the inclusion of polarization functions. The most recent study on the convergence of the SCF energy was reported by Kari and Csizmadia[96] who showed different limits attainable upon stepwise augmenting the basis set with p, d and f-type polarization functions. Absence of polarization functions in the wave function is also reflected in observables other than energy. Consider for example the dipole moment of the water molecule. The basis sets of the DZ size give the dipole moment round 9×10^{-30} Cm which is considerably higher than the value at the Hartree-Fock limit[8], 6.65×10^{-30} Cm; the remaining difference to experiment (6.19×10^{-30} Cm) is essentially due to the correlation energy[8]. The inadequacy of the DZ and any extended sp basis set in this case was demonstrated by Neumann and Moskowitz[57], who analyzed orbital contributions to the total dipole moment. It was shown[57] that the $1b_1$ contribution to the dipole moment must be zero in the DZ basis set on symmetry grounds. It becomes nonzero only if polarization functions are added to the basis set. Furthermore, the absence of any polarization in the $1b_1$ orbital results in a distortion of the charge distribution. This in turn has a large effect on the shape of the $3a_1$ orbital and its contribution to the dipole moment. Another example for such an analysis was communicated[97] for the inversion barrier of ammonia on which we shall comment in more detail in Section 5.C. For structures of certain types, the polarization functions are also important in geometry predictions, in spite of the fact that reliable bond lengths and bond angles are obtained with DZ or even minimum basis sets[88,98-101] for most molecules. Typical examples are H_3O^+, H_2O_2, SH_4 and ClF_4. Here correct structures are predicted only if polarization functions are included in the basis set. Thus, without polarization functions, H_3O^+ is predicted[100,102] to be planar instead of pyramidal, H_2O_2 to be in trans instead of nonplanar configuration[103,104], and SH_4, SF_4 and generally AB_4 molecules[105,106] to have C_{2v} instead of C_{4v} symmetry.

Some other information on the effect of polarization functions in chemical applications will be provided in Chapter 5. In this section we note the importance of polarization functions with particular first and second row atoms viz. the importance of 2p functions for Li and Be atoms[107,108], 3p functions for Na and Mg, and 3d functions for Na, Mg and the other second-row elements. For example, in developing a basis set for sodium of the DZ quality[109], 3p functions should not be omitted. According to Schaefer[99], the 3s-3p near degeneracy should be recognized and a 3p function added to the basis set. As regards 3d functions, they appear to be more important[101,110,111] for molecules containing second-row atoms than for molecules containing only first-row atoms. However the difference is rather quantitative than qualitative[112]. In some cases, the effect of d functions on the wave function of molecules with second row atoms is overestimated owing to the use of a minimum sp basis set[112]. Useful information on the relative importance of d functions with first row and second row atoms was provided by comparative studies such as e.g. on H_2O and H_2S[111] and NH_3 and PH_3[113]. Differing role of d functions was found in bonding in last two molecules. In PH_3 phosphorus d orbitals are found[113] to contribute directly to sigma bonding, whereas in NH_3 nitrogen d orbitals produce merely an angular effect which tends to reduce the calculated HNH angle to a value significantly closer to the experimental angle.

It should be noted that the total energy is a rather insensitive test of the quality of wave functions. Hence, if the importance of polarization functions is to be judged, one-electron first-order and

second-order properties, charge density contours, spectroscopic constants and some other molecular properties should also be tested.

Examine now the determination of exponents for polarization functions. Obviously, the atomic ground state calculations that are so
useful in the optimization of valence shell exponents cannot help us.
There is a possibility of performing calculations for excited states
of atoms. This approach is, however, not appropriate. The role of polarization functions is to polarize valence orbitals in bonds so that
the excited atomic orbitals are not very suitable for this purpose.
Chemically, more well-founded polarization functions are obtained by
direct exponent optimization in molecules. Actually, this was done for
a series of small molecules in both Slater and Gaussian basis sets.
Among the published papers, we cite[24,26-28,101,114-122]. Since exponent optimization for each molecule is unpractical, attempts were made
for a generalization of the determined optimum exponents for their
use with other molecules. A study of this kind was reported by Roos
and Siegbahn[111] who performed the d-exponent optimization for H_2O and
H_2S with the Gaussian double zeta basis set. The d-exponents for the
other first and second row atoms were obtained from the assumption
that the ratio $r_{max}(2p)/r_{max}(3d)$ is constant for the first row atoms
and $r_{max}(3p)/r_{max}(3d)$ for the second row atoms. The r_{max} values are
radii of maximal charge density for respective orbitals. The constants
for the two ratios were obtained from the oxygen and sulphur exponents
in H_2O and H_2S. The radial probability distribution function is given
by the product of the square of the radial part of the Gaussian function (2.10) and r^2. Upon differentiating this distribution function
with respect to r, it is easily found that the radius of maximal
charge density is given for the Gaussian (2.10) by

$$r_{max} = \sqrt{\frac{n_g}{2\zeta_g}} \qquad (2.20)$$

where n_g and ζ_g have the same meaning as in eqn. (2.10). For Slater
functions it holds

$$r_{max} = \frac{n}{\zeta_s} \qquad (2.21)$$

Relationships (2.20) and (2.21) may be also useful for interrelation

of exponents of STO's and GTF's, provided the latter are primitives
(which with polarization functions is a very common case). For example,
in H_2S the optimum STO exponent[115] ζ_{3d} is 1.708. The corresponding exponent for the GTF with the same radius of maximal charge density is
0.49, very close to the value, 0.54, obtained by a direct optimization
[111]. At least, the considerations noted in this paragraph may be useful for the determination of starting values in exponent optimization[116].

Pople and coworkers examined the effect of d-functions added to
the 6-31G basis set. For the resulting basis set, referred to as 6-31G*,
they suggested[107,108] the following α_d exponents: 0.2 for Li, 0.4 for
Be and 0.6 for B. The 6-31G* basis set was also applied to the series
of CH_4, NH_3, H_2O, HF, N_2 and C_2H_2 molecules for which the d-exponent
optimization was performed. Hariharan and Pople[117] concluded that the
optimum value of d-exponents does not depend strongly on chemical environment and suggested α_d = 0.8 as the standard value for all the
atoms C, N, O and F. Actually, this value does not differ significantly from the Gaussian d-exponents suggested by Dunning and Hay[121]:
0.7 for B; 0.75 for C; 0.80 for N; 0.85 for O; and 0.90 for F.

As regards the hydrogen 2p Gaussian exponent, a reasonable choice
[99,121] appears to be around 1.0. Optimum Slater exponent of the 2p
hydrogen function[99,114] ranges around 2.5, depending on the particular
molecule and the basis set. The hydrogen atom represents a special
case. If use is made of a single primitive Gaussian, there exists a
direct correspondence between STO and GTF exponents. It may be shown[38]
that if a Slater orbital (eqn. (2.3)) with a particular ζ_s and n_s is
to be approximated by a single primitive GTF, the optimum Gaussian exponent is given by

$$\gamma_{opt} = \left[\frac{(n_g - 1)!\ 2^{n_g}\ \sqrt{2}}{(2n_g - 3)!!\sqrt{\pi}} \ \frac{1}{4n_g + 4n_s(n_s - 1) - 1} \right]^2 \tag{2.22}$$

For the 2p function, relationship (2.22) gives γ_{opt} = 0.04527, which
corresponds to the STO exponent, ζ_s = $1/n_s$ = 0.5. On applying the
scaling relation (2.15), we get for ζ_s = 1.0 the GTF exponent, ζ_g =
$n_s^2 \gamma_{opt}$ = 0.18108 and for ζ_s = 2.53, which is a typical exponent[114] for
the polarization 2p STO, we get ζ_g = 2.53^2 x (2^2 x 0.04527) = 1.16.
Optimum exponents depend of course on a particular molecule and the
basis set on the other atoms. But for purposes of estimates of inter-

relation of STO and GTF exponents, the approach just noted may be used along same lines as considerations based on the radii of maximal charge density.

A thorough d-exponent optimization of Cartesian GTF's with respect to "chemical environment" of atoms in molecules was performed by the present authors with V. Kelló[123]. The exponents were optimized for the Dunning's DZ contracted Gaussian basis set[59], keeping the exponents of hydrogen p-functions fixed at α_p = 1.0. From Table 2.11 and Fig. 2.3

T a b l e 2.11
Exponents for d-type functions[123]

Valence state of atom	Suggested valence state exponent	Molecule treated and its optimum exponent	
$\diagup C \cdots$	0.80	CH_4	0.78
$(+)$ $-C \cdots$	0.70	CH_3^+	0.72
$-\dot{C} \cdots$	0.85	$CH_3^{\bullet}$	0.84
$\diagup C =$	0.70	H_2CO	0.70
$-C \equiv$	0.85	C_2H_2 (CO)	0.86 $(0.70)^a$
$N \equiv$	0.95	N_2	0.93
$\diagdown \bar{N} \diagdown$	0.55	NH_2^-	0.57
$\diagdown \dot{N} \diagdown$	0.70	$NH_2^{\bullet}$	0.71
$(+)$ $-N \cdots$	1.25	NH_3^+	1.25
$\diagdown N \cdots$	0.85	NH_3	0.83
$\diagup (+)$ $N \cdots$	1.20	NH_4^+	1.20
$(-)O-$	0.55	OH^-	0.57
$O =$	1.05	H_2CO (CO)	1.04 $(1.11)^a$
$\diagdown O \diagdown$	0.95	H_2O	0.92
$(+)$ $\diagdown O \cdots$	1.20	H_3O^+	1.22

a CO does not conform perfectly to any among the assumed valence states. Its optimum α_d were therefore disregarded in selecting va-lence state exponents.

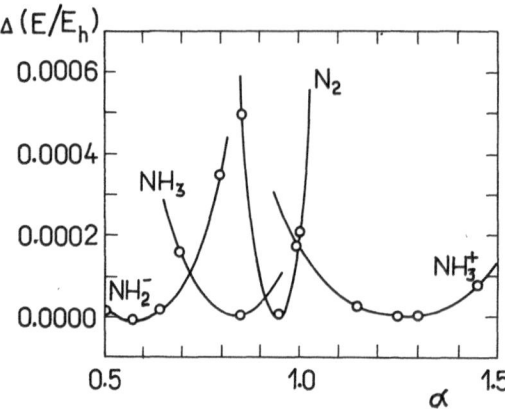

Figure 2.3

Dependence of the molecular SCF energy on the exponent of polarization function on the nitrogen atom. The energies are relative to the bottoms of the curves.

it is seen that the dependence of α_d on the molecular structure is not negligible. A large range of optimum α_d for different systems makes a choice of standard exponents for atoms difficult. Instead of atomic standard exponents we therefore suggested[123] standard exponents for particular valence states of atoms which we call "valence state" exponents. From the entries of Table 2.11 we attempted to formulate the following general rules:

proton addition - increase α_d by 0.35; H atom addition - use the same exponent; electron addition to the singly occupied orbital - lower α_d by 0.30.

The effect of the use of valence state exponents was tested on energies of reaction by making a comparison with the results obtained for for the standard[117] α_d = 0.8. From Table 2.12 it is seen that for most reactions the effect is small. For some reactions however it is significant and it may bring the calculated energies of reaction closer to Hartree-Fock values by more than 3-4 kJ/mol. Typical examples are the reactions involving species such as N_2 or CO_2, for which the total energy strongly depends on α_d (see Fig. 2.3). It should be noted that the optimum α_d exponents depend rather little on the underlying sp basis set, provided the latter is flexible enough. Therefore, the exponents given in Table 2.11 may be also used for other basis sets of similar size.

T a b l e 2.12

Energies of reactions ΔE (kJ/mol) and the effect of d-exponent optimization

Reaction	ΔE[a]			
	valence state α_d	$\alpha_d=0.8$	$\Delta\Delta E$[b]	$\exp(|\Delta\Delta E|/RT)$[c]
$NH_3^+ + NH_3 \rightleftharpoons NH_4^+ + NH_2$	−77.1	−77.1	0.0	1.0
$NH_3^+ + H_2O \rightleftharpoons H_3O^+ + NH_2$	+90.8	+90.7	+0.1	1.0
$NH_3 + H_3O^+ \rightleftharpoons NH_4^+ + H_2O$	−168.0	−167.8	−0.2	1.1
$NH_2 + H \rightleftharpoons NH_3$	−362.8	−363.1	+0.3	1.1
$NH_3^+ + H \rightleftharpoons NH_4^+$	−440.0	−440.3	+0.3	1.1
$C_2H_2 + 3H_2 \rightleftharpoons 2CH_4$	−495.2	−495.5	+0.3	1.1
$OH^- + NH_3 \rightleftharpoons H_2O + NH_2^-$	+67.5	+66.8	+0.7	1.3
$NH_2^- + H^+ \rightleftharpoons NH_3$	−1836.1	−1837.0	+0.9	1.4
$NH_2 + H^+ \rightleftharpoons NH_3^+$	−835.8	−834.4	−1.4	1.8
$NH_3 + H^+ \rightleftharpoons NH_4^+$	−912.9	−911.5	−1.4	1.8
$OH^- + H^+ \rightleftharpoons H_2O$	−1768.6	−1770.2	+1.6	1.9
$H_2CO + 2H_2 \rightleftharpoons CH_4 + H_2O$	−254.5	−256.8	+2.3	2.5
$H_3O^+ + OH^- \rightleftharpoons 2H_2O$	−1023.7	−1026.5	+2.8	3.1
$N_2 + 2H_2 \rightleftharpoons 2NH_2$	+197.5	+194.7	+2.8	3.1
$OH^- + NH_4^+ \rightleftharpoons H_2O + NH_3$	−855.8	−858.7	+2.9	3.2
$OH^- + NH_3^+ \rightleftharpoons H_2O + NH_2$	−932.9	−935.9	+3.0	3.4
$N_2 + 3H_2 \rightleftharpoons 2NH_3$	−171.3	−174.7	+3.4	3.9
$CO + CH_4 \rightleftharpoons C_2H_2 + H_2O$	+232.4	+228.5	+3.9	4.8
$CO + 3H_2 \rightleftharpoons CH_4 + H_2O$	−262.8	−266.9	+4.1	5.2

[a] The total energies for H_2 and H are −1.131197 and −0.497637 E_h respectively.

[b] This is the difference between the entries in the first two columns.

[c] This factor means the ratio of the theoretical equilibrium constants given by the two basis sets for T = 298 K.

2.G. Off-Centered Gaussian Functions

This section is devoted to Gaussian basis functions which are not placed on the atom centers but at different points in space. We note on the use of the following types of basis set functions:

   (i) Gaussian lobe functions

   (ii) Gaussian bond functions

   (iii) floating spherical Gaussian functions.

The development of the basis sets with off-centered functions aimed at avoiding use of higher spherical harmonics (i.e. d, f, ... type functions) without losing the basis set flexibility. Actually, the lobe function and floating spherical function basis sets are constructed only from s-type functions. Computationally, this restriction is very advantageous. The evaluation of integrals over s-type functions is very fast and the corresponding computer program may be simple. The Gaussian lobe function method was introduced by Preuss[124,125] and developed for routine calculations by Whitten[126]. The contraction of Gaussian lobe function (GLF) basis sets[49] is made along the same lines as with Cartesian GTF's. As regards the primitives, the s-type functions are expressed in the usual way. But the primitives of p, d, f, ... types are expressed as linear combinations of s-Gaussians (lobe functions) placed at different points so as to retain the proper symmetry (see Fig. 2.4). Thus, a p-type function on nucleus A may be

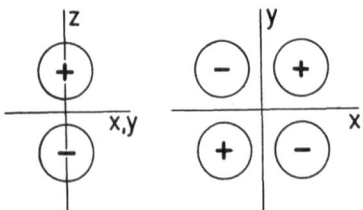

Figure 2.4
Representation of p_z and d_{xy} orbitals by lobe functions.

written as the difference between two lobe functions in the form

$$\chi_p = N(2\alpha/\pi)^{3/4}\left\{ \exp\left[-\alpha\,(r_A - R_0)^2\right] - \exp\left[-\alpha(r_A + R_0)^2\right]\right\} \quad (2.23)$$

where N is the normalizing factor for the p-function and $(2\alpha/\pi)^{3/4}$ is the normalizing factor for the lobe functions with the exponent α. R_0 means a constant displacement from nucleus A along the symmetry axis of the p orbital to be expanded. This displacement is usually defined as

$$R_o = C\alpha^{-1/2} \tag{2.24}$$

where C is the constant independent on α. By expanding the function (2.23) into the Taylor series it may be shown[127] that the GLF p-function approaches the Cartesian p-function in the limit $R_o \rightarrow 0$. On numerical grounds R_o cannot be set to an arbitrary small number. Typical choices for p functions are $C = 0.03$[127,128] and $C = 0.1$[129]. In fact, the quality of the approximation (2.23) depends on C rather slightly [127]. Anyway the representation of functions in Fig. 2.4 should be so understood that the lobes are heavily penetrating.

For the d and f functions, the most detailed discussion on the representation by lobe functions was reported by Driessler and Ahlrichs[130]. The functions d_{xy}, d_{xz}, d_{yz} and $d_{x^2-y^2}$ are constructed from four lobes as indicated in Fig. 2.4 by making use of eqn. (2.24) as with p-type functions. The representation of d_{z^2} functions is less straightforward[130,131]. Perhaps the most profitable approximation is the linear combination of three lobes[130]. One of them is placed on the atomic center and its exponent is scaled according to simple prescription[130] so that the analytical $(3z^2 - r^2)$ function is approached. The other two are shifted by $\pm R_o$ along the z direction by making use of eqn. (2.24). In this way, a complete set of five d-functions is represented by $(4 \times 4 + 3) = 19$ GLF's. For the representation of the f-set, 44 GLF's are required[130]. Recently a detailed study on GLF's was reported by Le Rouzo and Silvi[132].

It should be noted that in programs based on Cartesian GTF's use is mostly made of six d-functions: d_{xy}, d_{xz}, d_{yz}, d_{x^2}, d_{y^2} and d_{z^2}. However, the linear combination $d_{x^2} + d_{y^2} + d_{z^2}$ gives rise to a 3s function which has the same exponent as the d-functions. This redundant function may be eliminated in the SCF procedure by the basis set transformation to symmetry-adapted functions. The effect of the 3s function is very small at the SCF level, but as regards the correlation energy it may be larger[8]. Elimination of the 3s function is topical if the importance of d functions is to be examined or if comparison is to be made for calculations performed with lobe and Cartesian Gaussian functions. This applies particularly to small sp basis sets.

The SCF calculations on the second-row atoms with the lobe and Cartesian sp basis sets established the essential equivalence of the two approaches[128]. They also proved that exponents obtained for one type of Gaussian functions will work equally well for the other. Since that study[128] much experience has also been acquired with molecules.

Stated briefly, the two approaches provide practically the same results.

Another way of using off-centered functions is represented by the
addition of bond functions to the usual basis functions centered on
the nuclei. These are usually the simple s-type GTF's located on chemical bonds and lone pair orbitals; sometimes also p-type GTF's are
centered on the same place as the s-type bond functions. Again, the
first idea is due to Preuss[125]. The aim is to improve the electron
density around the chemical bonds and in electronic lone pairs upon a
moderate augmentation of the basis set. Ideally, the effect should be
the same as with the addition of a set of polarization functions, but
at a considerably lower cost. The utility of bond functions was tested
[133-145] by the calculations of various physical properties of molecules such as for example force constants[134,139], molecular geometries
and energy differences[140,142-144], dipole moments[143], and proton potential curves in hydrogen bonded systems[145]. Use of bond functions is
topical especially with large molecules, where the addition of a complete set of polarization functions is prohibitive. However, the bond
functions can hardly be expected to substitute fully the polarization
functions in applications in which the angular polarization is important. This is the case for one-electron properties; the quadrupole
moment of the nitrogen molecule may be taken as an example[141]. As regards the practical use of bond functions there is a drawback that
general rules for selecting exponents and positions of bond functions
are still lacking. It appears that the location of bond functions at
the midpoint of bonds is a reasonable choice[134,140]. Relevant contribution to this problem originated from the Theoretical Chemistry Laboratory of Vienna University[135,136]. From the results of the cited papers[135,136] it appears that the exponents as well as the positions of
bond Gaussian functions do not depend much on the specific molecule
but rather on the particular bond.

Another favorable finding was reported by Burton[143] that the exponents of bond functions optimized at the SCF level and with the inclusion of the correlation energy (by the CEPA method) are rather
close in absolute value and that bond functions are very effective in
accounting for correlation effects.

The last approach dealt with in this section is the Floating
Spherical Gaussian Orbital (FSGO) model and related methods. The original FSGO model introduced by Frost[146-148] is very simple. It treats
chemical bonds in terms of localized orbitals, in a very close correspondence with chemical concepts. Each localized orbital (inner

shell, bonding orbital or lone-pair orbital) is represented by means
of a floating spherical Gaussian function of the following form

$$\varphi_i = (2/\pi \rho_i^2)^{3/4} \exp\left[-(r - R_i)^2/\rho_i^2\right]$$ (2.25)

The parameters to be varied are the orbital radius ρ_i and the space
coordinates of the orbital center vector R_i. The orbital radius ρ_i is
related simply to the orbital exponent by

$$\alpha_i = \frac{1}{\rho_i^2}$$ (2.26)

The Slater determinant for a 2n-electron system is expressed by means
of n localized orbitals of the form (2.25). Since the model has only
nonlinear parameters, instead of the usual iterative SCF method the
direct search method is used. The technical details and the description of the computer program may be found in the original paper[147].
With the FSGO model rather numerous calculations have been reported
so that the virtues and shortcomings of the approach were well established[148-150]. The results obtained conform to what might be expected
for such a simple approach. Stated briefly[150], the FSGO model is a
useful alternative to the often less accurate semiempirical methods
or to the more time-consuming ab initio LCAO-MO procedures. A feeble
point of the FSGO approach is a poor representation of π-orbitals and
lone pairs. This is reflected, for example, in poor energy comparisons
between molecules with different numbers of multiple bonds[149,150] and
in predictions of bond angles[151] for molecules such as CH_2, OH_2 and
NH_3 which contain lone pairs. Among possible improvements of the original FSGO model the most natural one is the basis set extension by
describing each electron pair by means of a linear combination of
floating spherical orbitals[152-155],

$$\chi_i = N_i \sum_\mu C_{i\mu} \left(\frac{2\alpha_{i\mu}}{\pi}\right)^{3/4} \exp\left[-\alpha_{i\mu}(r - R_{i\mu})^2\right]$$ (2.27)

Parameters which are now to be optimized are all exponents $\alpha_{i\mu}$, orbital center position vectors $R_{i\mu}$ and linear coefficients $C_{i\mu}$. Ex-

tension of the basis set leads to a considerable improvement of calculated molecular geometries[152,153] and spectroscopic constants[153].
This is, however, associated with a rapid increase in the number of
nonlinear parameters which have to be optimized (C_i, α_i, R_i for each
added spherical lobe). It was therefore concluded[153] that for practical purposes the traditional SCF MO-LCAO approach· is preferable to
this general extension of the FSGO model. Obviously, if the advantages of the FSGO model are to be retained, some restrictions on the
expansion (2.27) must be imposed. In the so called "double Gaussian"
approach[154,155] a bond which is critical for the problem studied is
represented by two spherical lobes, but otherwise the wave function
is based on the original FSGO model. This approach is very effective.
For example, the simple FSGO method gives a barrier to rotation in
ethane of 23.8 kJ/mol. If a symmetrical pair of lobes is used to represent the C-C bonding orbital, the torsional barrier is improved[156]
to 12.9 kJ/mol, compared to the experimental value of 12.6 kJ/mol. In
our opinion the achieved results alone are not so important as the.
finding that the idea of FSGO permits various improvements and extensions[157,158] of the simple FSGO model. The most fruitful among them
is the molecular fragment model of Christoffersen and coworkers[154,
155]. In this method simple FSGO's of the form (2.25) are used as basis functions. Their parameters ρ_i and R_i are determined via energy
minimization calculations on molecular fragments such as e.g. CH_4,
·CH_3, NH_3, ·NH_2, H_2O and ·OH that are chosen to mimic the various
anticipated bonding environments. The fragments and their associated
FSGO's are combined appropriately to form the large molecule of interest, and an usual iterative SCF calculation is carried out, in contrast to the original FSGO approach. Each molecular orbital φ_i is
given by

$$\varphi_i = \sum_{A=1}^{P} \sum_{k=1}^{N_A} c_{ki}^A \, \chi_k^A \qquad (2.28)$$

where χ_k^A are either simple FSGO's or combinations of two FSGO's for
π-type orbitals. In the latter case the two FSGO's are placed symmet-
rically above and below the central atom, on a line perpendicular to
the plane of atoms, in analogy to Gaussian lobe functions. The sums are
taken over all fragments (P) and orbitals within a fragment (N_A). Parameters of FSGO's are used without modification in molecular calcula-

tions. This is a very advantageous feature of the molecular fragment approach because the optimization of nonlinear parameters in the FSGO model becomes troublesome for large molecules. Compared to current ab initio calculations, the number of basis set functions in the molecular fragment approach is smaller than it is in a standard minimum basis set approach. In spite of their small size, the basis set in the molecular fragment approach are highly flexible with respect to the bonding environment in the molecule. This flexibility originates from the fact that FSGO basis orbitals contain all GTF components, i.e. s, p, d, f, ... This may be shown[159] by the expansion of a FSGO in terms of atomic Gaussian functions. Moreover, from the analysis[160] of spatial transformation properties of FSGO basis sets it appears that full sets of p- and d-type atomic orbital components can be included in an FSGO basis using substantially fewer basis functions than needed in traditional lobe-functions basis sets. Hence, in spite of the simplicity the molecular fragment model is a method which permits acceptable accuracy to be achieved even with molecules as large as anthracene[161] and pentapeptides[162].

2.H. Comparison of Slater-Type and Contracted Gaussian Basis Sets

When comparing STO and CGTF basis sets of a particular size, it should be taken into account that with STO basis sets the results are only due to exponents of STO's, whereas with the CGTF basis sets also the effects of the number of primitives and their contractions are involved. Furthermore it should be kept in mind that the total energy is a rather insensitive test of the quality of wave functions, so that also some other molecular properties should be considered. Finally, a rigorous comparison should also include calculations going beyond the Hartree-Fock limit.

To start the discussion in a practical way, we comment first on the comparison of computational times involved. Unfortunately, there is a lack of data which give a direct evidence. To the best of our knowledge, the only information of this kind communicated so far is due to Hosteny and coworkers[7]. These authors reported the calculations on the water molecule performed with the STO DZ and CGTF DZ basis sets on the same computer. With the STO basis set, the computation time for the one- and two-electron integrals over atomic functions was ten times longer than with the CGTF basis set! Although the ratio depends obviously on the efficiencies of the respective programs, a significant difference in cost is generally to be expected. For other stages

of the calculation (also for CI with singly and doubly excited
states), the times were roughly comparable.

As regards the quality of results, some comparison was already
made in Table 2.10. The entries in Table 2.10 are typical in the following points: (i) a considerable range of energy with the minimal
basis sets; (ii) rather small energy differences with the DZ basis
sets; (iii) overestimation of dipole moments with optimized minimum
and DZ basis sets (good agreement with experiment for minimum STO basis sets is evidently fortuitous). Table 2.10 contains also the DZ
data of Hosteny and coworkers[7]: the total energy given by the CGTF
basis set is seen to be slightly lower than the STO energy. Table 2.13
suggests that this is a general rule rather than a special case for

T a b l e 2.13
Comparison of total energies (E/E_h) given
by STO DZ and CGTF DZ basis sets[a]

	CGTF[b]	STO	Ref.[c]
N_2	-108.8782	-108.8617	163
CO	-112.6850	-112.6755	56
NH_3	-56.1760	-56.1717	164
H_2CO	-113.8293	-113.8119	164

[a] For both basis sets the exponents are optimum for atoms. Essentially experimental
geometries used for all molecules.
[b] Taken from Ref. 123; Dunning's contraction[59] of the (9s5p/4s) set.
[c] References to STO calculations.

H_2O. A similar table might be also set up for minimal basis sets.
Here, however, a comparison would be of little use because of large
ranges of the total energy owing to its dependence on the choice of
STO exponents, number of Gaussian primitives, their exponents and the
way of contraction. Nevertheless also for minimal basis sets it is
possible to state that the CGTF basis sets yield typically lower energies than the STO basis sets. The CI calculations of Hosteny and coworkers[7] with the DZ basis sets permitted also the comparison of correlation energies given by the two basis sets. The valence shell correlation energy was larger with the CGTF basis set, whereas the inner
shell correlation energy was better accounted for by the STO basis set.

The fact that the minimum and DZ CGTF basis sets mostly give lower energies than the corresponding STO basis sets, need not be overemphasized. In actual chemical applications the absolute magnitude of the total energy need not be the most important criterion. For the purposes of comparison, it is more important to know the trends of calculated molecular properties on going to larger basis sets. Some information on this topic may be inferred from the results of calculations on N_2 collected in Table 2.14. Again, it is seen that with the [4s3p]

T a b l e   2.14

Comparison of CGTF and STO basis sets[a] for N_2

		Slater (4s3p)	Gaussian [4s3p]	Gaussian [4s3p2d]
Total energy	E_{SCF}	−108.8865	−108.8877	−108.9732
Second moments[b,c]	$\langle x^2 \rangle$	7.7443	7.5704	7.6061
	$\langle z^2 \rangle$	24.4246	24.3304	23.5666
Quadrupole moment[b]	Θ_{zz}	−1.7121	−1.7918	−0.9923
Potential[c]	$\langle 1/r_N \rangle$	21.6363	21.6367	21.6601
Electric field	$E_z(N)$	−0.1294	−0.2138	−0.0322
Electric field gradient	$q_{zz}(N)$	1.2433	1.2480	1.3574

[a] Refs. 59, 114. All quantities in relative values, see Appendix A.
[b] Relative to the center of mass.
[c] Electronic contribution.

basis set the CGTF energy is lower than the STO energy. Some one-electron properties given by the two basis sets are significantly different. The respective differences are however much smaller than in the case when either of the two basis sets is compared with the [4s3p2d] CGTF basis set containing polarization functions. Consider for example the quadrupole moment. The value given by the [4s3p2d] basis set is very close to the value of −0.9473 ea_o^2 obtained[98] at the Hartree-Fock limit and to the experimental value[165,166] of −1.1 ea_o^2. The two basis sets of the [4s3p] size yield quadrupole moments that are markedly different from these values. A more detailed analysis of the trends upon the basis set progression is permissible for the water molecule, for which extensive studies were performed[6,8,167]. By making use of a part of the tabular material from the cited papers and by supplementing it for data from other papers[168,169], it was possible

to set up comparative tables for the total energy (Table 2.15), one-
-electron properties (Table 2.16), calculated geometries and force
constants (Table 2.17) at both the SCF and CI levels. From the entries
in these tables it is possible to draw several conclusions, essentially in the form expressed by Rosenberg and coworkers[8,167]:

T a b l e   2.15

Comparison of CGTF and STO basis sets[a] for H_2O

Basis set		Energy			Dipole moment (SCF)
Primitive	CGTF or STO	SCF	CI-SD-FC[b]	CI-SD[c]	
1	(4s3p/2s)	-76.0200			9.26
2 (9s5p/4s)	[4s3p/2s]	-76.0109			8.95
3	(5s4p/3s)	-76.0238			8.99
4 (10s6p/5s)	[5s4p/3s]	-76.0207			9.05
5	(4s3p1d/2s1p)	-76.0596	-76.2690	-76.2879	7.16
6 (9s5p2d/4s1p)	[4s3p1d/2s1p]	-76.0495	-76.2541	-76.2704	6.87
7 (11s7p1d/6s1p)	[4s3p1d/2s1p]	-76.0554			
8	(5s4p1d/3s1p)	-76.0632	-76.2777	-76.3163	6.85
9 (11s7p1d/6s1p)	[5s4p1d/3s1p]	-76.0520	-76.2665	-76.2826	
10	(5s4p2d/3s1p)	-76.0642	-76.2990	-76.3398	6.65
11 (10s6p2d/4s1p)	[5s4p2d/2s1p]	-76.0589	-76.2885	-76.3232	6.87
12 (13s8p3d1f/ 6s2p1d)	[8s5p3d1f/4s2p1d]	-76.0659			
Estimated limit[d]		-76.0675 ± 0.0010	-76.374 ± 0.006	-76.4382 ± 0.0024	6.19[e]

[a] Energies in relative values E/E_h, dipole moments in 10^{-30} Cm. Calculations 1-4 from Ref. 6; 5, 6, 8, 10, 11 from Ref. 8; 7 and 12
from Ref. 168; 9 from Ref. 169.
[b] CI including the SCF function and all single and double excitations from it, except that the la_1 orbital is constrained to be
doubly occupied in all configurations (frozen core).
[c] CI including the SCF function and all single and double excitations from it.
[d] Ref. 8.
[e] Experimental value[87].

(i) In contrast to minimum and DZ basis sets, the SCF energy given
by a larger STO basis set is always lower than that given by the CGTF
basis set of the same size (in terms of contracted groups). A plausible explanation is the role of the number of primitive Gaussians. Actually in larger CGTF basis sets, the number of primitives is usually
less than the double of the number of STO's. As documented by calcula-

T a b l e 2.16

Comparison of some one-electron properties of H_2O calculated with STO and CGTF basis sets[8]

	STO (5s4p2d/3s1p)		CGTF [5s4p2d/2s1p]		
	SCF	CI-SD	SCF	CI-SD	Experiment
Dipole moment (10^{-30} Cm)					
μ	6.6549	6.4021	6.8681	6.6443	6.1863 ± 0.002
Quadrupole moment tensor[a] (10^{-40} cm^2)					
θ_{xx}	-8.419	-8.396	-8.326	-8.316	-8.34 ± 0.07
θ_{yy}	8.736	8.706	8.633	8.616	8.77 ± 0.07
θ_{zz}	-0.317	-0.310	-0.307	-0.300	-0.43 ± 0.10
Diamagnetic susceptibility tensor[a,b,c] (10^{-12} m^3/mole)					
χ^d_{av}	-15.262	-15.497	-15.239	-15.487	-14.6 ± 2.0
$\chi^d_{xx} - \chi^d_{av}$	-0.954	-0.959	-0.971	-0.973	-1.06 ± 0.01
$\chi^d_{yy} - \chi^d_{av}$	0.886	0.891	0.904	0.907	0.95 ± 0.01
Average diamagnetic shielding at the oxygen and hydrogen nuclei (in ppm)					
σ^d_{av} (O)	416.09	416.28	416.07	416.10	~414.6
σ^d_{av} (H)	102.42	102.33	102.39	102.31	102.4
^{17}O quadrupole coupling tensor[b,d] eqQ/h (MHz)					
xx component	11.58	10.92	11.50	10.79	10.17 ± 0.07
yy component	-10.37	-9.71	-10.28	-9.58	-8.89 ± 0.03
$(q_{zz} - q_{yy})/q_{xx}$	0.791	0.779	0.789	0.775	0.75 ± 0.01

[a] Relative to the centre of mass.
[b] Only two components given; the third can be determined from the zero trace condition.
[c] The average susceptibility χ^d_{av} and the anisotropies $\chi^d_{gg} - \chi^d_{av}$ (g = x,y) are given.
[d] Computed from the electric field gradient q at the oxygen nucleus, using the value Q = -0.0867 x 10^{-40} cm^2 for the ^{17}O nuclear quadrupole moment.

tions 6 and 7 (in Table 2.15), the extension of the primitive set may be important. Some effect may also be assigned to the extent of expo-

T a b l e 2.17

Comparison of geometries and force constants calculated for H_2O with STO and CGTF basis sets[167,169]

Property	STO (5s4p2d/3s1p)			CGTF [5s4p1d/3s1p]		Experiment[c]
	SCF	CI-SD[a]	CI-SDQ[b]	SCF	CI-SD-FC[a]	
$r_e (10^{-10}$ m)	0.9398	0.9527	0.9573	0.944	0.960	0.957
α (degrees)	106.08	104.93	104.58	105.3	103.8	104.5
$f_{rr} (10^2$ N/m)	9.7934	8.8757	8.539	9.50	8.44	8.45
$f_{rrr} (10^2$ N/m)	−10.36	−9.70	−9.50	−10.00	−9.84	−9.55
$f_{\alpha\alpha}/r_e^2 (10^2$ N/m)	0.8751	0.8135	0.801	0.816	0.752	0.76

[a] See footnotes in Table 2.15.
[b] CI-SD with the estimated contribution of quadruple excitations.
[c] For references see Ref. 167.

nent optimization in the two basis sets. Note that the lowest SCF energy for H_2O obtained so far was attained with the CGTF basis set[168].

(ii) The two basis sets give almost the same valence-shell correlation energy. Its magnitude is dependent predominantly on the number of virtual orbitals. The inner-shell correlation energy is significantly larger with the STO basis sets.

(iii) In the calculation 10 (see Table 2.15) the SCF energy approached the Hartree-Fock limit to within 0.003 E_h. The convergence of the correlation energy towards the exact value is considerably slower. For the same basis set only 75% of the correlation energy is reproduced, the difference from the "exact" value being 0.098 E_h.

(iv) One-electron properties, molecular geometries and force constants given by the two types of large basis sets differ rather little. If correlation energy is accounted for, which is important in some cases, excellent agreement of calculated physical properties with experiment is found. The only significant discrepancy encountered with the water properties concerns the deuteron quadrupole tensor, which implies a deficiency in the description of the wave function in the neighbourhood of the hydrogen atoms[8]. In contrast, in some cases such as χ_{av}^d and $\sigma_{av}^d(0)$, the computed property values are presumably more accurate than the experimental data. Also the statistical uncertainties in the cubic and diagonal quartic force constants[167], not presented in Table 2.17, are considerably smaller for the theoretical

than for the experimental values. A thorough examination of the potential energy hypersurface and spectroscopic constants of the water molecule was reported recently by Hennig et al.[170] who used a very extended CGTF basis set [7s5p2d/4s2p] at the SCF, CI-SD and CI-SDQ levels. Their data are essentially in agreement with previous results
that are presented partly in Table 2.17.

From the results an important conclusion follows viz. that for
large basis sets the differences in actual basis sets are not reflected
significantly in the computed molecular properties. It is just this
feature which is typical for ab initio calculations.

To summarize the discussion, the Gaussian and Slater basis sets of
the same size and quality give comparable energies and the other molecular properties.

2.I. Remarks on the Selection of the Basis Set

This section contains a few practical remarks on the compatibility
of the basis set selected with the problem to be investigated. As late
as in 1971, the situation was characterized by Rothenberg and Schaefer
[133]: "the selection of basis functions for ab initio molecular calculations still seems to be more an art than a science". Soon after this
time considerable experience with basis set selection was accumulated.
Although for a particular problem it is hardly possible to give a suggestion for an unequivocal basis set selection, a few general rules
may be formulated that facilitate the situation considerably.

First of all, when selecting a basis set for a particular problem,
some attention must be paid to what Mulliken[171] calls a "good balance"
of the basis set. A well-balanced basis set ensures that any atomic
(or molecular) property is described with the relative error which is
roughly equal for all valence orbitals. For the first-row atoms, the
problem is equivalent to finding the right balance between the number
of s- and p-type GTF's and to decide how many s-type GTF's one should
place on a hydrogen atom in conjugation with s/p sets in molecular calculations. According to van Duijneveldt[67], the s/p ratio of the following primitive sets is too low
        2/1 3/2 4/3 6/4 8/5 9/6 10/7 12/8
whereas that of the following is too high
        4/1 5/2 7/3 9/4 11/5 12/6 13/7 15/8
If the size of the basis set for a first-row atom is (nsmp) then the
use of m or m+1 s-type GTF's on H is recommendable. Generally, the
equally balanced set for polyatomic molecules means "poor atomic set

for all atoms, medium set for all atoms, good atomic set for all the atoms in the molecule[168]".

Usually, the exponents of hydrogen s-functions in molecules are scaled by a factor of 1.2 for STO basis sets. This corresponds to the multiplication of the GTF exponents by a factor[56,59] of 1.44. For the other atoms best atomic exponents are mostly taken, at least for DZ and larger basis sets.

Polarization functions should be added only to good atomic basis sets (DZ or better). Otherwise the polarization functions not only introduce the polarization effects but also substitute the effect of that part of the sp set which is missing in the atomic basis set[168].

The general rules noted above make of course not the problem of the basis set selection solved. In some cases the use of basis sets of nonstandard composition is required. We comment on some of them. First, the rules conform to neutral and positively charged species. For negatively charged species the basis set extension is recommended[121, 171-178]. For molecules of the first-row atoms this means to augment the basis set with diffuse p-type functions, in some cases possibly also with s-type functions on heavy atoms and hydrogen atoms. Dunning and Hay[121] recommend augmentation of the (9s5p) GTF set with the p-type functions having the exponents: B (0.019); C (0.034); N (0.048); O (0.059) and F (0.074). If the usual contraction to the DZ set is employed, the resulting basis set for the first-row atoms is of the [4s3p/2s] size. The effect of the addition of diffuse p-functions is very large. It brings the computed electron affinities[121] and energies of reaction containing negative ions[179,180] closer to experiment by as much as 40 kJ/mole. The effect of diffuse p-functions cannot be substituted by the addition of polarization functions to standard sp basis sets. Diffuse functions are also topical with the transition metal compounds. As shown by Hay[181] a set of d-functions optimum for the ground state of a transition metal element should be augmented by a single diffuse 3d function. This accounts for the fact that electronic configurations of transition metals in their ground states and compounds are different (the atomic ground states are mostly $4s^2 3d^{n-2}$ whereas in molecules the $4s3d^{n-1}$ and $3d^n$ configurations are usually met) and that 3d orbitals become more radially diffuse in the sequence $4s^2 3d^{n-2}$, $4s3d^{n-1}$, $3d^n$.

Another type of applications, in which diffuse function must be used, are calculations on electronic spectra. With Rydberg states, the importance of diffuse functions is obvious. By their electronic structure the Rydberg states resemble the corresponding molecular cation plus a weakly bonded electron which is housed in a molecular orbital formed predominantly by atomic orbitals with higher than valence shell principal quantum numbers, i.e. by 3s, 3p, 4s, ... orbitals with the first-row atoms. For exponents of these diffuse functions use may be made of appropriate Slater exponents which may be transformed to Gaussian exponents as noted with polarization functions in Section 2.F. Dunning and Hay[121] suggested a set of single GTF's optimized for the Rydberg states of the first- and second-row atoms. If a higher basis set flexibility is required, they suggested a standard split to two diffuse functions. It should be noted that for the representation of Rydberg states the diffuse functions need not be necessarily placed at the atomic centers. The nature of Rydberg states permits placing them [182-185] at the center of charge of the limiting ion state or simply at the center of the molecule. In some cases it is necessary to augment the basis set with diffuse functions also in treatments of valence-excited states since the latter may contain a considerable admixture of Rydberg states[186]. This is topical for the regions of avoided crossing of two states, one being a valence and the other being a Rydberg state. Among many known examples, we point out avoided crossing of $^2\Sigma^+$ states in OH[187] and CF[188] radicals. Nevertheless also some excited states that were believed to be entirely valence in character on the basis of semiempirical calculations are substantially diffuse. A striking example are the lowest $\pi \rightarrow \pi^*$ singlet states in ethylene and butadiene. Ab initio calculations overestimate these $\pi \rightarrow \pi^*$ transitions drastically, unless the basis set is augmented with diffuse functions, in which case a very good agreement with experiment is achieved[189,190].

A special selection of the basis set is also necessary for calculations of polarizabilities[191,192]. A meaningful calculation requires the use of a large flexible basis set containing diffuse d-functions (and also the inclusion of correlation effects). An interesting type of basis sets oriented to calculations of polarizabilities was proposed by Sadlej[193,194]. In his approach use is made of an electric-field-variant Gaussian basis set, i.e. a basis set explicitly dependent on the strength of the external electric field perturbation. This approach permits polarizabilities to be obtained without polarization functions, provided the basis set contains additional p-type Gaussians with rather low orbital exponents.

For some other molecular properties the wave function must be accurate in the proximity of nuclei. Such a property is for example the hyperfine splitting caused by Fermi contact term which is proportional to the total spin density at the nucleus. Among the one-electron properties, it is the electric field gradient (and the derived deuteron quadrupole coupling tensor) which is very sensitive with regard to the quality of the wave function near the nucleus, especially at the hydrogen[8].

For the sake of compactness, we embodied the knowledge about the basis set selection in Table 2.18. This Table provides of course only

T a b l e 2.18
Compatibility of the basis set with the problem treated

Problem	Lowest appropriate size of the basis set	Comment
Molecular geometry	Minimum	Except for dihedral angles and pyramidal structures where use of polarization functions is needed
Force constants	DZ	Mostly reasonable agreement with experiment
Rotational barriers	Minimum	For H_2O_2 and some other molecules use of DZ+P sets is required
Inversion barriers	DZ+P	Polarization functions are very important
Energies of reaction	DZ	For semiquantitative energy predictions
	DZ+P	Accurate predictions; for negative ions diffuse functions must be added
One-electron properties	DZ+P	Agreement with experiment for minimum basis sets is fortuitous
Polarizability	Large	Two sets of polarization functions, one with diffuse functions ($\alpha_d \approx 0.1$)
Electronic spectra	Minimum, DZ	Diffuse functions for Rydberg states
Fine structure parameters	Minimum, DZ	Good representation of inner shells
Interaction of ions and dipoles, hydrogen bonded systems	Minimum	Appropriate for relative stabilities and geometries; counterpoise correction should be tested
Weak intermolecular interactions	Large	Diffuse polarization functions in addition to common polarization functions; counterpoise correction should be tested

a rough guide. More detailed information on the basis set selection may be found with particular topics in Chapter 5 and papers cited there. Table 2.18 refers to SCF calculations. It should be kept in mind that in calculations with inclusion of the correlation energy, the choice of the basis set is governed by different rules. Generally, if correlation effects are to be included, use of a set smaller than the DZ+P basis set should be avoided unless the approach at the SCF level is completely meaningless and the use of such a large basis set is prohibitive.

3. SCF Calculations

This chapter is devoted to SCF calculations which represent the overwhelming majority of the reported ab initio calculations. We are not going here to treat general problems of SCF theory which are also inherent in popular semiempirical methods such as PPP and CNDO. These features will be intentionally suppressed. Instead, emphasis will be laid on specific problems of the ab initio SCF approach that are not encountered in semiempirical MO methods and on the progress achieved in solution of these problems in last years.

As with semiempirical methods, the problem to be solved is given by the Hartree-Fock-Roothaan (HFR) equations

$$\sum_\nu (F_{\mu\nu} - \varepsilon_i S_{\mu\nu}) c_{i\nu} = 0 \tag{3.1}$$

where $F_{\mu\nu}$ are matrix elements of the Hartree-Fock operator, ε_i orbital energies, $S_{\mu\nu}$ overlap integrals, and $c_{i\nu}$ are coefficients of the LCAO expansion (2.1). For closed-shell systems it holds[195]

$$F_{\mu\nu} = H_{\mu\nu} + G_{\mu\nu} \tag{3.2}$$

$$F_{\mu\nu} = H_{\mu\nu} + \sum_\lambda \sum_\sigma D_{\lambda\sigma} \left[(\mu\nu|\lambda\sigma) - \frac{1}{2} (\mu\lambda|\nu\sigma) \right] \tag{3.3}$$

Generalization to open-shell systems does not represent any specific problem of ab initio calculations and therefore it will not be treated here. In eqn. (3.3), $H_{\mu\nu}$ denotes the matrix element of the one-electron part of Hamiltonian, $D_{\mu\nu}$ is the element of the density matrix

$$D_{\mu\nu} = 2 \sum_i^{occ} c_{i\mu} c_{i\nu} \tag{3.4}$$

and $(\mu\nu|\lambda\sigma)$ are two-electron integrals

$$(\mu\nu|\lambda\sigma) = \iint \chi_\mu(1)\chi_\nu(1) \frac{1}{r_{12}} \chi_\lambda(2)\chi_\sigma(2) \, d\tau_1 d\tau_2 \tag{3.5}$$

In semiempirical methods of the PPP and CNDO types, the $H_{\mu\nu}$ elements are approximated by expressions containing empirical parameters. In ab initio treatments they are of course calculated rigorously. Since the one-electron part of Hamiltonian (1.1) contains two terms, the $H_{\mu\nu}$ elements are composed of two contributions

$$H_{\mu\nu} = T_{\mu\nu} + V_{\mu\nu} \tag{3.6}$$

which are due to kinetic energy and nuclear attraction operators. Hence besides the overlap and electronic repulsion integrals, it is also necessary to compute in ab initio calculations the $T_{\mu\nu}$ and $V_{\mu\nu}$ integrals:

$$T_{\mu\nu} = - \int \chi_{\mu}(1) \frac{1}{2} \nabla^2(1) \chi_{\nu}(1) \ d\tau_1 \tag{3.7}$$

$$V_{\mu\nu} = - \int \chi_{\mu}(1) \sum_{C} \frac{Z_C}{r_{1C}} \chi_{\nu}(1) \ d\tau_1 \tag{3.8}$$

Ab initio calculations also differ considerably from semiempirical treatments in the second term of the $F_{\mu\nu}$ element (3.2) which contains the summation over the electronic repulsion integrals. Most of the latter are neglected in semiempirical treatments, whereas in ab initio calculations they are all considered and computed accurately unless they are vanishing or smaller than an a priori chosen threshold (see Section 3.C.).

At the ab initio level, the total electronic energy is given by

$$E_{el} = \sum_{\mu} \sum_{\nu} D_{\mu\nu} H_{\mu\nu} + \frac{1}{2} \sum_{\mu} \sum_{\nu} D_{\mu\nu} \sum_{\lambda} \sum_{\sigma} D_{\lambda\sigma} \left[(\mu\nu|\lambda\sigma) - \frac{1}{2} (\mu\lambda|\nu\sigma) \right] \tag{3.9}$$

The total SCF energy is obtained by adding the nuclear repulsion energy

$$E = E_{el} + \sum_{A > B} \frac{Z_A Z_B}{R_{AB}} \tag{3.10}$$

As with semiempirical methods, the system of HFR equations (3.1) is solved iteratively. The experience acquired with semiempirical treatments may be very profitable here because many problems are common to both semiempirical and ab initio calculations. We mean for example convergence of the SCF procedure, guess of the starting eigenvectors, convergence hastening and damping of oscillations. There are, however, some specific problems of ab initio calculations that are not encountered in semiempirical calculations. The most important among them is a problem of rapid evaluation of integrals appearing in eqn. (3.2), in particular, of two-electron integrals (3.5). It should be realized that for a basis set of n functions the number of two-electron integrals is of the order n^4. This is an essential difference from semiempirical methods where the number of integrals is proportional to N^2 (for the PPP method N is a number of atoms involved in the π-electronic system, for CNDO it is the number of valence-shell atomic orbitals). Even for rather large molecules, the N^2 integrals may be stored in a computer core during the whole run. This is not possible with ab initio calculations. The integrals are stored on a magnetic tape or disk. Thus, in addition to the limiting factor of computer time we are to face another problem, namely a problem of data handling and data retrieval. In Chapter 2 we discussed the reduction in the number of integrals by means of contraction of Gaussian basis sets. This facilitates manipulation of integrals and saves computer time. As shown by Clementi[196,197] continuous gains are also due to improvements in the numerical analysis, programming techniques and organization, and in making more efficient use of increasing performance and utilities of computers. In this chapter we comment on the most important procedures that save computer time in the SCF calculations. The two consecutive steps of the SCF calculations, evaluation of integrals and solution of the HFR equations (3.1), are discussed separately. The first step appears to be still more important because it is the most time-consuming. However, prior to discussing the possibilities of reducing computer time for evaluation of integrals, it is profitable to note briefly on the way how the integrals are computed (for more specific information, see special reviews[47,198]).

3.A. Integrals over Slater-Type Orbitals

In early days of ab initio calculations, the situation with regard to developing effective machine subroutines for the rapid evaluation of integrals over STO's seemed hopeful. According to Parr[9], "a 1951 conference promised them shortly[199]; in 1958 there was an announcement concerning some of them[200,201]. But, in 1962, a letter appeared in the Communications column of the Journal of Chemical Physics describing a new method for obtaining the worst of them, which ought to work[202]. At the present time (i.e. 1963), in fact, computer routines for two-center integrals involving 1s, 2s, and 2p orbitals are routinely available in several laboratories, and routines for two--center integrals for 3p, 3d, etc. are about to become available. Three- and four-center integrals are much harder to come by, although Boys is systematically producing them[203]".

This documents that in spite of a great effort the solution of the problem proceeded slowly. The first effective computer programs for routine calculations were restricted to diatomic and later also to linear polyatomic molecules. For polyatomic molecules of any geometry several computational methods have been developed. It appears that the most frequently used among them are the Gaussian transform method of Shavitt and Karplus[202,204] and the ζ-function method of Barnett[205,206]. The most recent review on this topic was reported by Saunders[198] and the most widely used program appears to be POLYCAL, QCPE 161, written by Stevens[29].

3.B. Integrals over Gaussian-Type Functions

GTF's possess an important property namely that the product of two GTF's having different centers A and B is itself a Gaussian (apart from a constant factor) with a center P somewhere on the line segment AB. This property simplifies considerably the computation of integrals because it reduces two-center integrals to one-center integrals and four-center integrals to two-center integrals. Specifically, for s--type GTF's the following holds

$$e^{-\alpha r_A^2} \cdot e^{-\beta r_B^2} = K e^{-\gamma r_P^2} \tag{3.11}$$

where K is a constant

$$K = e^{-\frac{\alpha\beta}{\alpha+\beta} \overline{AB}^2} \tag{3.12}$$

$$\overline{AB}^2 = (A_x - B_x)^2 + (A_y - B_y)^2 + (A_z - B_z)^2 \tag{3.13}$$

$$\gamma = \alpha + \beta \tag{3.14}$$

$$P_x = \frac{\alpha A_x + \beta B_x}{\alpha + \beta} \tag{3.15}$$

and similarly for P_y and P_z. When the GTF's are not 1s orbitals, an extra factor such as $x_A^{\ell_1} x_B^{\ell_2}$ appears in their product. According to the Saunders' notation[198]

$$x_A^{\ell_1} x_B^{\ell_2} = \sum_{k=0}^{\ell_1+\ell_2} f_k(\ell_1,\ell_2,\overline{PA}_x,\overline{PB}_x) x_P^k \tag{3.16}$$

where

$$\overline{PA}_x = P_x - A_x \tag{3.17}$$

$$x_P = x - P_x \tag{3.18}$$

$$f_k(\ell_1,\ell_2,\overline{PA}_x,\overline{PB}_x) = \sum_{i=0}^{\ell_1} \sum_{j=0}^{\ell_2} {}^{i+j=k} \, \overline{PA}_x^{\ell_1-i} \binom{\ell_1}{i} \overline{PB}_x^{\ell_2-j} \binom{\ell_2}{j} \tag{3.19}$$

where the summations expand over the indices i and j with the restriction i+j=k. Analytical formulas for integrals over s-type GTF's were derived first by Boys[31]. For the purposes of the forthcoming discussion it is useful to present here his formulas for the unnormalized functions

$$(\chi_A|\chi_B) = \left(\frac{\pi}{\alpha+\beta}\right)^{3/2} e^{-\frac{\alpha\beta}{\alpha+\beta}\overline{AB}^2} \tag{3.20}$$

$$(\chi_A|T|\chi_B) = \frac{\alpha\beta}{\alpha+\beta}\left(3 - \frac{2\alpha\beta}{\alpha+\beta}\overline{AB}^2\right)\left(\frac{\pi}{\alpha+\beta}\right)^{3/2} e^{-\frac{\alpha\beta}{\alpha+\beta}\overline{AB}^2} \tag{3.21}$$

$$(\chi_A | V_C | \chi_B) = \frac{2\pi}{\alpha + \beta} F_o \left[(\alpha + \beta)\overline{CP}^2 \right] e^{-\frac{\alpha\beta}{\alpha+\beta} \overline{AB}^2} \tag{3.22}$$

$$(\chi_A\chi_B | \chi_C\chi_D) = \frac{2\pi^{5/2}}{(\alpha + \beta)(\gamma + \delta)\sqrt{\alpha + \beta + \gamma + \delta}}$$

$$\times F_o \left[\frac{(\alpha + \beta)(\gamma + \delta)}{\alpha + \beta + \gamma + \delta} \overline{PQ}^2 \right] e^{-\frac{\alpha\beta}{\alpha+\beta} \overline{AB}^2 - \frac{\gamma\delta}{\gamma+\delta} \overline{CD}^2} \tag{3.23}$$

In this formulas, χ_A, χ_B, χ_C, χ_D are, respectively, GTF's with expo= nents α, β, γ, δ. The points P and Q lie on $\overline{AB}$ and $\overline{CD}$, respective- ly, and are given by eqn. (3.15). F_o is defined by

$$F_o(x) = \frac{1}{2}\sqrt{\frac{\pi}{x}} \, erf(\sqrt{x}) \,, \qquad x > 0 \tag{3.24}$$

where erf is the error function.

From the formulae it is seen that the computation of integrals is rather simple. The most time-consuming step is the evaluation of the F_o function. In modern programs the values of F_o are obtained by means of tabular interpolation[198]. In the original Boy's approach the inte- grals for higher quantum numbers were derived from basic formulas (3.20)-(3.23) by differentiation with respect to parameters A_x, A_y, etc. At the present time the integrals are calculated in another way which was described by Taketa and coworkers[50].

We do not present here formulas for integrals over GTF's with higher azimuthal quantum numbers. These may be found in a recent pa- per by Saunders[198] where also an analysis is given for the effective integral computation.

3.C. Computer Time Saving in Evaluation of Integrals

Discussion in this section is focussed on the computation of inte- grals with contracted Gaussian basis sets, though many approaches noted here are of general importance and may be applied also to STO basis sets. This holds especially for storing a part of the data obtained in a particular run and having it available for later use in subsequent runs. Very efficient data handling of this kind is provided for by making use of routines "merge", "more", "add" and "substract" involved in the program IBMOL-4[207]. These routines may be easily introduced into other programs, though modern programs mostly have these facilities. To illustrate the utility of such a data handling, consider for example the potential curve for the motion of hydrogen in the hydrogen bond of the water dimer. For a DZ basis set the total number of basis functions in the dimer is 28 which means that it is necessary to compute 82621 two-electron integrals (no use is assumed of the plane of symmetry of the system). If the calculation is repeated for a new geometry, which differs from the original one only by the position of the proton in the hydrogen bond, the two-electron integrals for 26 functions preserve their original value and may therefore be used from the original list of integrals. Their number is 61776. Hence it is sufficient to recalculate only 20845 integrals which means approximately a three quarters time-saving. The calculation noted is of course rather modest. In calculations on very large systems, such as e.g. the calculation on the hydrogen bond in the system cytosine-guanine[208], the time-saving is considerably higher. Equally, one can apply this procedure to basis set optimization in which the exponent is varied for one or several functions, to cases in which the size of the system treated is to be enlarged or reduced, and to cases in which the calculation is to be repeated for a larger or smaller basis set. The data storing is also desirable for having the possibility of restarting a run at any point after the interruption.

Another way of reducing the time of integral evaluation is based on the use of symmetry. Probably the most ingenious use of molecular symmetry, both local and that of the point group, is made in the POLYATOM[209,210] program. Here prior to the proper calculation those integrals are eliminated which are zero due to the symmetry properties of basis set functions or which may be derived in absolute value from others by means of any symmetry operation. Only so called "unique" nonvanishing integrals are computed. Next integrals, which are related to the unique integral by the symmetry properties of basis functions and may differ from it only by a sign, are stored consecutively after it. The integrals that can be derived from $(\mu\nu|\lambda\sigma)$ by permutation of indices

$$(\mu\nu|\lambda\sigma) = (\mu\nu|\sigma\lambda) = (\nu\mu|\lambda\sigma) = (\nu\mu|\sigma\lambda) = (\lambda\sigma|\mu\nu) = (\lambda\sigma|\nu\mu) = (\sigma\lambda|\mu\nu) = (\sigma\lambda|\nu\mu)$$

are not tested on symmetry and they are neither computed nor put in
the integral list. Accordingly, they are referred to as equivalent in-
tegrals. Usually, $\mu \geq \nu, \lambda \geq \sigma$, $(\mu\nu) \geq (\lambda\sigma)$ where $(\mu\nu) = \mu(\mu - 1)/2 +$
ν . The efficiency of the use of symmetry may be demonstrated with a-
cetylene. For a DZ basis it is necessary to compute 45150 two-electron
integrals. If use is made of the D_{4h} symmetry, which is the highest
symmetry of acetylene in the representation of Cartesian GTF's, their
number reduces to 6558.

The two treatments just discussed refer to manipulation with integrals over contracted Gaussian or Slater functions. They have a common feature - the improvement is achieved by means of program organization. Let us now pass from the data handling to the computation of
integrals itself, but restricting ourselves to contracted Gaussian basis sets. It will be shown that also here much may be gained by the
effective organization of the program. The discussion will be based on
the paper by Ahlrichs[211]. Assume contracted GTF's that are expressed
by linear combinations of primitive Gaussians centered in the point I
as follows

$$\chi_\mu = \sum_i c_i g_i (\alpha_i, I, s, t, u) \tag{3.25}$$

Assume next for the sake of simplicity that χ are of s-type, so that
$s = t = u = 0$ and the normalized primitives in eqn. (3.25) become

$$g_i(r) = \left(\frac{2\alpha_i}{\pi}\right)^{3/4} e^{-\alpha_i(r-r_i)^2} = \left(\frac{2\alpha_i}{\pi}\right)^{3/4} e^{-\alpha_i r_I^2} \tag{3.26}$$

A two-electron integral $(\mu\nu|\lambda\sigma)$ may be written by using eqn. (3.25) as

$$(\mu\nu|\lambda\sigma) = \sum_{i,j,k,\ell} c_i c_j c_k c_\ell (g_i g_j | g_k g_\ell) \tag{3.27}$$

The direct calculation of $(\mu\nu|\lambda\sigma)$ over the summation (3.27) is inef-
fective because it involves redundant recalculation of many partial
terms. This may be best seen if eqn. (3.27) is rewritten in the following form (compare (3.23) and take into account normalization of
primitives (3.20)):

$$(\mu\nu|\lambda\sigma) = \sum_{k,\ell} t_{k\ell} \sum_{i,j} t_{ij} B^{1/2} F\left[B(\overline{PQ}^2)\right] \qquad (3.28)$$

where

$$t_{ij} = c_i c_j \int g_i g_j \, d\tau = c_i c_j (4\alpha_i \alpha_j)^{3/4} (\alpha_i + \alpha_j)^{-3/2} \, e^{-\frac{\alpha_i \alpha_j}{\alpha_i + \alpha_j} \overline{IJ}^2} \qquad (3.29)$$

$$P_x = \frac{\alpha_i I_x + \alpha_j J_x}{\alpha_i + \alpha_j} \qquad (3.30)$$

$$Q_x = \frac{\alpha_k K_x + \alpha_\ell L_x}{\alpha_k + \alpha_\ell} \qquad (3.31)$$

$$B = (q_{ij} + q_{k\ell})^{-1} \qquad (3.32)$$

$$q_{ij} = (\alpha_i + \alpha_j)^{-1} \qquad (3.33)$$

and where $F = 2/\sqrt{\pi}\, F_o$, F_o being given by eqn. (3.24), and $\overline{IJ}^2$ and $\overline{PQ}^2$ are defined as in eqn. (3.13). From eqns. (3.28)-(3.33) it is seen that a fast integral evaluation would be easy if the quantities t_{ij}, P_x, P_y, P_z and q_{ij} could be kept in storage for all i and j. This means, however, large requirements on the computer core used. Since the calculation of t_{ij} is the most time consuming, it is profitable to keep in storage at least t_{ij} values. This is feasible for common basis set with the number of primitives, say, n = 150. Another pos-sibility is to divide t_{ij} and q_{ij} into blocks of appropriate length and keep only those blocks which are currently needed in storage. This principle of preserving and utilizing information common to sev-eral integrals also permits effective computation of integrals over p,d ... functions, although in a somewhat different way. As we learn-ed in Section 3.B., the product of two primitive Gaussians can be ex-pressed as a linear combination of primitive Gaussians centered at the same point. Hence, if any of the two functions is of other than s-type, then additional terms appear in the expression for the product (see eqns. (3.16) and (3.19)). These terms represent again information

common to many integrals. Provided the same exponents are used for a
set of three p-functions p_x, p_y, p_z, which is the most usual case,
making use of the common information facilitates evaluation of as many
as $3^4 = 81$ integrals and with a six-component d set as much as $6^4 = 1296$ integrals. This idea is used very efficiently in the program[212]
Gaussian 70. This program was developed for basis sets discussed in
Section 2.E. that are typical of a "shell structure". The constraint
imposed on the exponents of valence-shell orbitals ns, np_x, np_y, np_z
to share a common value permits utilization of information common to
a shell of basis functions; for a sp shell this concerns $4^4 = 256$ different integrals over primitive functions. A more detailed information
may be found in the paper by Pople and Hehre[213].

Comparison of various procedures, discussion on their merits and
drawbacks, and typical timings of current methods were reviewed by
Dupuis and coworkers[214].

Consider now another way of accelerating the calculation of integrals, a way of "controlled numerical approximation". Its application
was prompted by the observation that in large molecules many integrals
$(\mu\nu|\lambda\sigma)$ are almost vanishing[197,211,215]. Neglect of integràls smaller
than a chosen threshold manifests itself as a small error in the computed SCF energy. This error may be estimated a priori. After some
numerical experimentation it is possible to acquire experience about
the magnitude of this error in its dependence on the chosen threshold,
so that the accuracy of the SCF calculation may be defined a priori.
It should be realized that one works with a finite number of digits
anyway, so that this neglect of integrals represents by no means any
approximation but a procedure fully controlled numerically. Effective
use of this approach requires, however, a fast estimate of the magni-
tude of $(\mu\nu|\lambda\sigma)$ integrals, or contributions to them $(g_i g_j | g_k g_\ell)$ over
primitives, prior to their calculation. In order to arrive at a con-
venient integral test, let us continue in the analysis by Ahlrichs[211].
For the terms B and F(x) noted above the following inequalities hold
(for the properties of the F function, see ref. 47,198):

$$B = (q_{ij} + q_{k\ell})^{-1} \leqq (4q_{ij}q_{k\ell})^{-1/2} \tag{3.34}$$

$$F(x) \leqq 2\pi^{-1/2} \tag{3.35}$$

This permits us to estimate rigorously the contribution over primitives to the $(\mu\nu|\lambda\sigma)$ integral. By introducing the quantities

$$u_{ij} = t_{ij}q_{ij}^{-1/4} \tag{3.36}$$

it follows from expressions (3.34) and (3.35)

$$|t_{ij}t_{k\ell}B^{1/2}F(B(\overline{PQ}^2))| \leqq (2/\pi)^{1/2}u_{ij}u_{k\ell} \tag{3.37}$$

This estimate permits neglect of all contributions to the integral $(\mu\nu|\lambda\sigma)$ for which $u_{ij}u_{k\ell} < T$ holds, T being a chosen threshold. Significant time saving is achieved with this procedure in very large molecules or in molecular systems consisting of fragments sufficiently separated. For such molecules the dependence of the computer time on the basis set size is not as drastic as n^4. Some typical timings are given in the cited papers[196,211].

The details of the testing procedure depend on the particular program used. We present the utility of the method by means of the calculations performed with the program[209] POLYATOM/2. The problem treated is the H_2-H_2 system calculated for different intersystem distances. A treatment of such a small system is somewhat untypical, but it is useful for our purposes because it shows clearly how the computation time is related to the intersystem (or "inter-fragment") distance. The basis set used was [4s3p]. For a linear structure of the complex this basis set contained 28 contracted functions which were constructed from 36 primitives[216]. Two thresholds were used: $T = 10^{-11}$ and $T = 10^{-8}$ E_h. The first one ensures practically the full accuracy and it can be safely applied even to weak interactions. The second one ensures the accuracy still better than to seven decimal places. This conforms to the experience[196,211] that the error in the SCF energy is smaller than 10 T. The entries in Table 3.1 indicate what time can be saved for large molecules. From the foregoing discussion it may be concluded that the integral time is not a simple function of the number of contracted and primitive functions only but also of a chosen threshold, interatomic distances, exponents of primitives and the molecular symmetry.

Considerable reduction of the computer time may be also achieved by numerically controlled procedures of another type, in which the integrals smaller than a chosen threshold T' are calculated approx-

T a b l e 3.1

Dependence of the number of computed two-electron integrals on the intermolecular distance in the H_2-H_2 system

Distance[a] (r/a_o)	cut-off T = 10^{-11} E_h		cut-off T = 10^{-8} E_h	
	nonequivalent integrals	unique integrals[b]	nonequivalent integrals	unique integrals[b]
3.0	81786	41041	81260	40771
4.0	80620	40449	79118	39691
6.0	74853	37546	69459	34838
8.0	64660	32430	58088	29132
10.0	55227	27697	47725	23933
15.0	36150	18132	29674	14890
20.0	24805	12455	21921	11013

[a] Intermolecular distance between the centers of the two H_2 molecules; linear configuration assumed.
[b] Number of unique integrals, actually computed; use of symmetry was made (center of inversion).

imately. The threshold is so chosen that the pertinent integrals are small enough but not negligible. This approach is used in the "adjoined technique" which was introduced by Clementi[196,197]. It is based on the replacement of a contracted function by a single Gaussian function (adjoined function). If any integral $(\mu\nu|\lambda\sigma)$ over contracted functions is lower than T', it is not computed accurately but a value is assigned to it of the integral $(\mu'\nu'|\lambda'\sigma')$ over adjoined functions μ', ν', λ', σ'. Consider for example that each contracted function in $(\mu\nu|\lambda\sigma)$ contains five primitives. In that case the $(\mu\nu|\lambda\sigma)$ integral is built up from 625 integrals over primitive functions. Obviously, avoiding this by the calculation of a single integral over primitives brings about considerable time saving. The exponent of the adjoined Gaussian is so chosen as to give maximum overlap of the adjoined function with the contracted GTF. For a system cytosine-guanine[196], the introduction of the adjoined basis set resulted in the reduction of the integral time by a factor of 100. The method was later discussed by Ahlrichs[211] who showed that an adjoined function should in general be constructed from two Gaussians. This is a probable reason why the adjoined basis sets of this type are not currently used.

Nevertheless the development of approximate calculations of small integrals is still in progress[217,218]. A very promising approach is based on the concept of approximate charge densities. According to Whitten[217] it may be assumed that charge densities in the integral $(\chi_\mu(1)\,\chi_\nu(1)|\,\chi_\lambda(2)\,\chi_\sigma(2))$ will be approximated by $\chi_\mu(1)\,\chi_\nu(1) \simeq a_{\mu\nu}\Phi_{\mu\nu}(1)$ $= \Phi'_{\mu\nu}(1)$ and $\chi_\lambda(2)\,\chi_\sigma(2) \simeq a_{\lambda\sigma}\Phi_{\lambda\sigma}(2) = \Phi'_{\lambda\sigma}(2)$, where $\Phi_{\mu\nu}$ are functions, containing variational parameters which are optimized together with $a_{\mu\nu}$ parameters for all $n/2(n+1)$ product densities. For example, taking the product density approximation as $\chi_\mu(1)\chi_\nu(1) \simeq a_{\mu\nu}\chi'_\mu(1)\chi'_\nu(1)$ (and analogously for $\chi_\lambda(2)\,\chi_\sigma(2)$) gives, after the parameter optimization, the integral approximation

$$(\chi_\mu(1)\chi_\nu(1)|\,\chi_\lambda(2)\,\chi_\sigma(2)) \simeq a_{\mu\nu}\,a_{\lambda\sigma}\,(\chi'_\mu(1)\,\chi'_\nu(1)|\chi'_\lambda(2)\,\chi'_\sigma(2)) \tag{3.38}$$

Here χ'_μ is a contracted GTF which, with respect to χ_μ, contains a smaller number of primitives, or it simply contains a single primitive GTF. Even for the latter case, this approximation is entirely different from the integral approximation using a single Gaussian which has a maximum overlap with the contracted function; $a_{\mu\nu}$ factors exist, and Gaussian exponents when optimized do not necessarily lead to maximum overlap. With a suitable choice of $\Phi'_{\mu\nu}$, the method may also be used for a Slater basis set, in which case the time saving of integral evaluation is due to the possibility of reducing four-center integrals to two-center integrals[217].

For the approximations treated in this section it is possible to define rigorously and to estimate safely the error limits, so that the SCF energy can be kept within an a priori determined accuracy range. Consequently, the nature of the "accurate" ab initio calculation is preserved in spite of the neglect of approximate evaluation of some integrals. In contrast, there are also approaches, which may be called semi-ab initio methods, where the simplifications are not so rigorously justified. These methods are, however, beyond the scope of this book.

3.D. Computer Time Saving in the SCF Procedure

As we have seen in the preceding section, considerable progress was achieved in the integral evaluation. Less attention was paid to hastening the SCF procedure itself. This is understandable because for current basis sets and molecules the integral time represents usually a dominant portion of the total time, unless some use is made of facilities and tricks noted in the preceding section. In such a

situation any effort for hastening the SCF procedure would not result
in a significant time saving. Nowadays this does not hold any longer,
in particular for large molecules. Therefore attempts at a better program organization and the search for new effective procedures are
highly topical. An excellent and comprehensive review on this subject
was presented recently by Veillard[219]. Here we note briefly on possible improvements in the following points: (i) choice of the initial
density matrix; (ii) construction of the Fock matrix; (iii) diagonalization methods; (iv) convergence hastening. The points (i), (iii) and
(iv) do not represent problems specific for ab initio calculations,
but they should be noted for the sake of completness. A degree of success achieved in solution of problems (i) and (iv) is reflected in the
number of necessary iterations. In some cases it even decides whether
the SCF procedure converges at all. Time per one iteration is under
control of points (ii) and (iii). The matrix diagonalization does not
represent any serious problem because modern effective methods are
available and the time involved in the matrix diagonalization rises at
most as n^3. So most attention is paid to the construction of the Fock
matrix (eqn. (3.2)), where the fourth power dependency on the size of
basis set is encountered.

Let us start the discussion with the points (i) and (iv). As with
semiempirical methods of the PPP and CNDO types, also with ab initio
SCF calculations one starts with a Hückel-type calculation. In practice this is made by making the matrix density in expression (3.3)
equal to zero. Thus, initial guess at a set of MO coefficients is
formed by diagonalizing the one electron part of the Fock matrix. This
approach always risks a slow convergence or, even worse, sometimes the
SCF procedure does not converge at all, the oscillations in the total
SCF energy being the most typical case. A possible way of avoiding
poor convergence is the use of localized orbitals as starting wave
functions[220]. In the cases tested by Letcher[221] the calculation proceeded quickly to completion without a divergence. To the same conclusion arrived Shipman and Christoffersen[222] who used localized orbitals in the framework of FSGO's. Much time may also be saved if for
the initial guess use is made of the eigen vectors of an isoelectronic system or of eigen vectors given by a smaller basis set. For example it is very typical that the system treated is first calculated
with a DZ basis set and then with a DZ+P basis set. In such a case it
is profitable to start the DZ+P calculation with the DZ eigen vectors
supplemented with zero expansion coefficients for polarization functions. In calculations of interaction energies of molecules, the con-

vergence of a supersystem calculation is considerably improved if orthogonalized vectors of subsystems are used for starting wave functions. Similarly, in optimizations of molecular geometries and exponents of basis functions, it is very efficient to start the calculation with the density matrix given by a previous run for another geometry or a basis set with some different exponents.

Time saving in the SCF procedure may be also achieved in other ways than by a suitable guess of the starting wave function. Mostly, the programs are equipped with extrapolation procedures (for a general discussion see ref. 21) that estimate a new density matrix from three consecutive iterations and use it in the next iteration. Sometimes the total SCF energy oscillates in spite of all precautions. In such a case it is recommendable to attempt a damping procedure which for the density matrix D assumes an average from the density matrices D_1 and D_2 of the two successive iterations, $D = aD_1 + bD_2$, a and b being arbitrary parameters with the constraint $a + b = 1$. The problem of selecting a and b was discussed recently[223].

Great difficulties with convergency are often encountered with open shell systems for which even a combination of the techniques noted may be helpless. Such cases are called intrinsically divergent. Among methods which are applicable to these cases we note on the level shifting technique suggested by Guest and Saunders[224]. The method was developed for the improvement of convergence of the Roothaan RHF procedure[225] as a generalization of the level shifting technique for closed shell systems[226]. Its essence lies in computing the Fock matrix in the basis of the approximate Hartree-Fock orbitals from the previous iteration and adding a sufficiently large positive shift parameter to the open-shell diagonal elements and a larger parameter to the virtual-space diagonal elements. Properly chosen shift parameters guarantee that the first order contributions to the energy are negative.

The crucial problem in the point (ii) - the construction of the Fock matrix - is a retrieval and effective manipulation of the integrals $(\mu\nu|\lambda\sigma)$. Since all nonvanishing integrals must be read in from the tape or disk in each iteration, this manipulation may be very time consuming. Programs greatly differ on this point. From the discussion [211,219,227-229] on this topic it follows that it is profitable to have a record of integrals in a certain order which contains a sequence of three integrals $(\mu\nu|\lambda\sigma)$, $(\mu\lambda|\nu\sigma)$, $(\mu\sigma|\nu\lambda)$. This is required by an effective evaluation of the second term in eqn. (3.3). This term may be rewritten as

$$G_{\mu\nu} = \sum_{\lambda\sigma} D'_{\lambda\sigma} \left\{ \left[(\mu\nu|\lambda\sigma) - \frac{1}{4}(\mu\lambda|\nu\sigma) - \frac{1}{4}(\mu\sigma|\nu\lambda) \right] (1 - \frac{1}{2}\delta_{\mu\nu,\lambda\sigma}) \right\} \quad (3.39)$$

where D' is the density matrix in which the off-diagonal elements are doubled

$$D'_{\mu\nu} = 4 \sum_{i}^{occ} c_{i\mu} c_{i\nu} \left(1 - \frac{1}{2}\delta_{\mu,\nu} \right) \quad (3.40)$$

and $\delta_{\mu\nu,\lambda\sigma}$ is zero, unless $(\mu\nu) = (\lambda\sigma)$, $(\mu\nu) = \mu(\mu - 1)/2 + \nu$, in which case it is unity. The whole expression standing in (3.39) after $D'_{\mu\nu}$ may be taken as the element of the supermatrix P,

$$P_{\mu\nu,\lambda\sigma} = \left[(\mu\nu|\lambda\sigma) - \frac{1}{4}(\mu\lambda|\nu\sigma) - \frac{1}{4}(\mu\sigma|\nu\lambda) \right] \left(1 - \frac{1}{2}\delta_{\mu\nu,\lambda\sigma} \right) \quad (3.41)$$

The matrix elements (3.41) may be constructed if all three integrals are available at the same time. To achieve this, it is most profitable to compute them just in the order $(\mu\nu|\lambda\sigma)$, $(\mu\lambda|\nu\sigma)$, $(\mu\sigma|\nu\lambda)$. Other-wise, they must be reordered. The matrix elements $P_{\mu\nu,\lambda\sigma}$ are then stored on a tape or disk together with the pair indices $(\mu\nu)$, $(\lambda\sigma)$. They contribute to two G-matrix elements as follows:

$G_{\mu\nu}$ formed from contributions $P_{\mu\nu,\lambda\sigma} \times D'_{\lambda\sigma}$

$G_{\lambda\sigma}$ formed from contributions $P_{\mu\nu,\lambda\sigma} \times D'_{\mu\nu}$

(3.42)

The proper construction of the G matrix elements with respect to per-mutations of μ, ν, λ, σ indices in P and D' is ensured by the terms $(1 - 1/2\delta)$ in the definitions of the P and D' matrices (note that $D'_{\mu\nu} = D'_{\nu\mu}$ and $P_{\mu\nu,\lambda\sigma} = P_{\nu\mu,\lambda\sigma} = P_{\mu\nu,\sigma\lambda} = P_{\lambda\sigma,\mu\nu}$). Only the nonequiva-lent P and D' elements with $\mu \geq \nu \geq \lambda \geq \sigma$ are computed and stored.

Use of $P_{\mu\nu,\lambda\sigma}$ terms represents an alternative way of constructing the Fock matrix. In the traditional way, one uses directly the inte-grals $(\mu\nu|\lambda\sigma)$ listed in any arbitrary order. This case occurs with the POLYATOM program if use is made of symmetry relationships between ba-sis functions. As is usual (see p. 60), from eight equivalent integral list. The permutation of indices in the construction of the Fock matrix is then recognized in such a way that each integral contributes to one to six F matrix elements, depending on the integral type. Thus, an integral of the type $(\mu\mu|\mu\mu)$ contributes only to one element, whereas $(\mu\nu|\lambda\sigma)$ integrals with four distinct indices contribute to six elements. Any of the six F matrix elements, $F_{\mu\nu}$, $F_{\lambda\sigma}$, $F_{\nu\sigma}$, $F_{\nu\lambda}$, $F_{\mu\sigma}$, $F_{\mu\lambda}$, is formed (see eqn. (3.3)) from the products of integrals $(\mu\nu|\lambda\sigma)$ and the corresponding density matrix elements. Since the P terms are formed only once before starting the SCF procedure, the disadvantage of the traditional integral handling becomes obvious: as the number of iterations is increased, time saving with the use of $P_{\mu\nu,\lambda\sigma}$ terms approaches the factor of three.

Finally, we comment briefly on the use of symmetry. By making use of the symmetry point group of the molecule, the Fock matrix may be transformed into a blocked form and the eigenvalue problem (3.1) is then solved block by block. Transformation into a blocked form may be performed in different ways which will be not discussed here. We note only that the gain is not very important since the time involved in diagonalization rises at most as n^3. In some cases, however, the use of the symmetry-blocked F matrix may eliminate convergence problems. More significant time saving might be expected if symmetry were also used in the construction of the F matrix elements which is a n^4 dependent process. Several authors[230-236] have paid attention to this problem. The task lies essentially in finding an efficient use of symmetry properties of integrals over basis set functions. As we learned in Section 3.C., the integral package generated by POLYATOM/2 contains sequences of symmetry-related integrals from which only each first was computed. Winter and coworkers[231] developed such a procedure that also in the SCF procedure only the first integral from a group of symmetry-related integrals is needed. This permits the integral list to be reduced and consequently facilitates its processing in each iteration. The idea of this approach is based on the theorem that symmetry-related integrals over basis set functions make equal contributions to symmetry-adapted integrals with totally symmetric integrands[234]. The algorithm which deletes symmetry-related integrals from the integral list may also be applied[227] to the approach of constructing the F matrix by means of P integrals (3.41). Time saving is proportional to the integral list reduction which may be significant[231] for molecules of high symmetry. A recent detailed discussion on the use of molecular symmetry was reported by Dupuis and King[236].

In this chapter we have outlined the main problems of the SCF ab

initio calculations. Some of them are known from semiempirical calculations, the others are specific for ab initio calculations. Most of
the latter are related to the drawback of the n^4 dependence. We have
shown some tricks by means of which consequences of this bottleneck of
ab initio calculations may be reduced and we have indicated a progress
achieved in this field in the several last years. Beyond doubt the
methodological development will continue. Together with a continuous
development and availability of effective computers, this gives reason
for an optimistic outlook for a widespread applicability of ab initio
calculations to larger molecules. Nevertheless even at the present
stage the ab initio SCF calculations may be considered to be more or
less routine.

4. Correlation Energy[+]

Until recently the notion "ab initio calculations" was mostly understood as "ab initio SCF calculations". As a matter of fact, prior to 1970 the calculations of correlated wave functions for polyatomic molecules were rather exclusive. The situation changed dramatically in the early seventies. Effective algorithms were developed that yield highly correlated wave functions, some of them at a cost only moderately higher than that required for the respective SCF run. So it is perhaps not too optimistic to state that also the calculations with the inclusion of (a part of) the correlation energy are becoming routine. In this chapter we attempt to survey the present state of the art. In Section 4.B. a distinction is made between the problems, where the role of the correlation energy is small and those where it is of crucial importance. As regards the calculations beyond the Hartree-Fock limit there is a large variety of computational methods. We selected those that, in our opinion, are the most suitable for practical purposes or that yield perspectives for further development. The theoretical backround of the methods is intentionally suppressed. Instead, emphasis is laid on the fundamental idea on which the theory is based, the numerical feasibility, cost, the portion of the correlation energy recovered and reliability in chemical applications. The whole chapter is oriented to ground states.

4.A. Definition and Origin of the Correlation Energy

Within the framework of the SCF-MO approach, the probability density of finding simultaneously two electrons with different spin in a certain space is given simply by a product of probability densities of the individual electrons. This independence of electrons with different spin is, however, physically unrealistic, because the $1/r_{12}$ term in the Hamiltonian imposes a certain constraint, i.e., a certain "correlation" on the motion of all electrons in the system. Disregarding the electron correlation brings about the energy difference between the exact and SCF solutions which is called the correlation energy (Fig. 4.1). Here by the exact solution we imply the lowest energy attainable with the Hamiltonian (1.1) by a method which furnishes the

---

[+] This chapter is based partly on the reviews of I. Hubač with one of the authors[237,238].

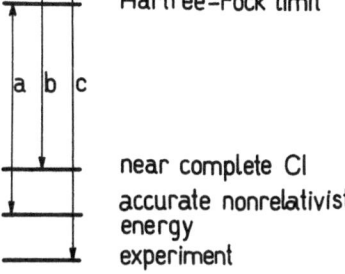

SCF with the minimum basis set

near Hartree-Fock
Hartree-Fock limit

a b c

near complete CI
accurate nonrelativistic
energy
experiment

Figure 4.1
Energy scale and the correlation energy.
a – correlation energy according to the definition; b – correlation energy estimated in practice; c – "experimental" correlation energy.

"upper bound" of energy and by the SCF solution we mean the lowest energy attainable within the Hartree-Fock approximation, i.e. the Hartree-Fock limit. From Fig. 4.1 it follows that the correlation energies may be estimated rather than determined. As a matter of fact, apart from a few very simple systems, neither the Hartree-Fock limits nor the exact nonrelativistic energies are accessible by ab initio calculations. With the former the situation is somewhat better because for a series of small molecules very extensive calculations were reported and the Hartree-Fock limits were estimated that are believed to be typically within $0.002\ E_h$ or less. To approach the true nonrelativistic energy is more difficult. Actually, for triatomic and larger molecules more than 80% of the correlation energy is only rarely recovered, even if most ingenious methods are used and considerable computational effort is exerted. For this reason the so called "experimental" correlation energies (see Fig. 4.1) are sometimes used. However, their reliability is questionable, too. The adoption of the Born-–Oppenheimer approximation and the neglect of relativistic effects appear to be justifiable (see Introduction), but the determined experimental correlation energies are again affected by the basis set limitation and, even worse, by the uncertainties in the experimentally determined heats of formation for many unstable species.

As regards the magnitude, the relative value of the correlation energy is small. Compared to the SCF energy, it amounts approximately to 1%. In absolute value, however, the correlation energy is large. Even with small molecules it is by one or more orders higher than heats of reactions and energies of activation of chemical processes.

Nevertheless we shall see in the following section that the situation is not so hopeless as it might appear.

4.B. Conservation of the Correlation Energy

In the field of ab initio calculations there exists an apparent controversy. On the one hand, neglect of the correlation energy is demonstrated theoretically as a crude approximation of the Hartree--Fock treatment and on the other hand, the majority of calculations is performed just at the level of the Hartree-Fock approximation and mostly with meaningful results. The fact that correlation effects may be disregarded in many cases has been recognized a long time ago. The tendency of cancellation of correlation effects in some chemical processes have been noted by several authors in the early sixties[239]. The conditions for this conservation of correlation energy may be expressed[240] as follows: 1. The number of electron pairs must be conserved. 2. Also the spatial arrangement must be approximately maintained for electron pairs which are nearest neighbors.

The first condition is satisfied automatically with all reactions containing closed shell molecules only. A systematic examination for this type of reactions was performed by Snyder and Basch[241,242]. The theoretical (SCF) heats of reactions were claimed to be more accurate than those obtained using semiempirical relations of bond energies for reactions of strained molecules, or those not well represented by a single valence-bond structure. However, Snyder and Basch concluded [241,242] that if a level of chemical accuracy is to be approached, some semiquantitative prediction of the change in correlation energy is required. Obviously, the actual conservation of correlation energy depends on the extent to which the second condition is satisfied. With respect to this condition the following classification[243] can be made for reactions with closed shell molecules:

Homodesmotic reactions. These are reactions[244] in which (i) there are equal numbers of bonds of a particular type (e.g., C[4]-C[4], C[4]-C[3], C[3]-C[3], C[3]=C[3], where the numbers in brackets indicate the total numbers of other atoms bonded to each carbon atom, so that C[4]-C[4] and C[3]=C[3] may represent CC bonds in ethane and ethylene, respectively) and (ii) there are equal numbers of each atomic type (such as C[4], C[3], etc.) with zero, one, two and three hydrogen atoms attached in reactants and products. The following examples[244-246] may be given:

$$CH_3CH_2CH_2CH_3 + CH_3CH_3 \longrightarrow 2\ CH_3CH_2CH_3 \qquad (4.1)$$

$$CH_3CH_2C \equiv CH + CH_3CH_3 \longrightarrow CH_3CH_2CH_3 + CH_3C \equiv CH \qquad (4.2)$$

$$RO^- + CH_3OH \longrightarrow CH_3O^- + ROH \quad (R = alkyl) \qquad (4.3)$$

$$RCC^- + HCCH \longrightarrow HCC^- + RCCH \qquad (4.4)$$

$$(4.5)$$

$$(4.6)$$

In homodesmotic reactions the structural elements in reactants and products match closely so that only a small change in correlation energy is to be expected. Indeed, for processes of the type (4.1) and (4.2) the heats of reactions given by DZ and DZ+P (often even by minimum) basis sets reproduce[244] experiment typically to within 4-9 kJ/mol. An ideal example of a homodesmotic process is the internal rotation in ethane and it is therefore not surprising that the SCF calculations give the respective rotational barrier in excellent agreement with experiment.

Isodesmic reactions. This concept introduced by Hehre and collaborators[247] means the reactions in which there is retention of the number of bonds of a given formal type, but with a change in their relation to one another. The homodesmotic processes are actually a subclass of isodesmic reactions. Typical representants of isodesmic reactions are so called "bond separation" reactions[247] such as for example

$$CH_3CHO + CH_4 \longrightarrow CH_3CH_3 + H_2CO \qquad (4.7)$$

$$CH_3OCH_3 + H_2O \longrightarrow 2\ CH_3OH \qquad (4.8)$$

Here a degree of the conservation of spatial relation depends evidently on the particular case. With simple reactions such as (4.7) and
(4.8), the error in computed heats of reactions[244,247] is larger than
it is with homodesmotic reactions, the typical value being 4-17 kJ/mol.

Anisodesmic reactions. Unfortunately most reactions of chemical
interest belong to this category of processes that are not isodesmic
and for which still larger correlation effects may be expected. As
might be noticed, homodesmotic and isodesmic reactions are mostly ar-
tificial processes of little interest to chemists. One can, however,
combine them in thermochemical cycles and to arrive in this way at
real chemical problems[248,249]. An example of the anisodesmic reaction
with a large change in correlation energy is provided by the dimerization $2BH_3 \longrightarrow B_2H_6$. The SCF calculation[250] gives for the energy of
dimerization the value of 87 kJ/mol. If the correlation energy is included (by the CEPA method; see Section 4.I.), one arrives at the value[250] of 153 kJ/mol. The number of electron pairs is preserved in this
reaction but their spatial arrangement and, which is more important,
the number of pair-pair interactions are different. On the other hand,
the reaction $NH_3 + HCl \longrightarrow NH_4Cl$ may also be taken as an anisodesmic
process. Here, however, the two conditions for the conservation of
correlation energy are well satisfied, so it was possible to predict
gaseous NH_4Cl by mere SCF calculations[251]. Existence of NH_4Cl was
later established experimentally[252].

For the correlation energy to be roughly conserved, not all reaction components need necessarily be closed shell species. For example,
for the process $BH_2 \longrightarrow BH + H$ a rather small change in correlation energy was predicted[253]: the UHF calculation gives for the dissociation
energy 340.2 kJ/mole, whereas the perturbation calculation up to third
order gives 351.9 kJ/mole. Although this process involves a bond fission, it does not involve the formation or rupture of an electron pair

$$\text{(4.9)}$$

This also applies[254] to a general class of A: + B• interactions leading to diatomic systems, such as for example

$$Na(^2S) + He(^1S) \longrightarrow NaHe(^2\Sigma^+) \tag{4.10}$$

and

$$\text{He}(^1\text{S}) + \text{F}(^2\text{P}) \longrightarrow \text{HeF}(^2\Pi) \tag{4.11}$$

As the opposite to the examples given above, we note now processes that involve a fission of electron-pair bonds. Here the change in cor-relation energy is extremely large and the Hartree-Fock approximation is inherently incapable of giving a reasonable account of heats of re-action. A very illustrative example is provided by potential curves of diatomic molecules. From Fig. 4.2 it is seen that for larger depar-

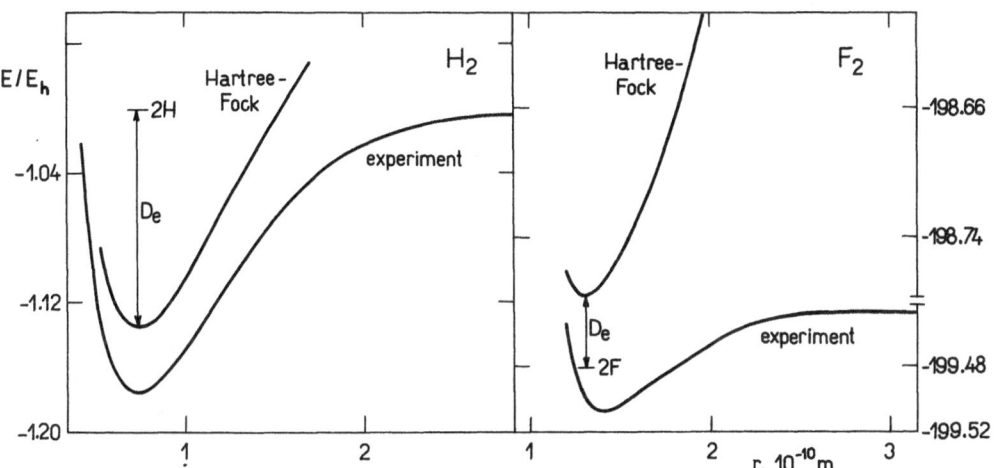

Figure 4.2
Potential curves for H_2 and F_2 (plotted from tabulated data in Ref. 255). The dissociation energies (indicated by arrows) are determined from Hartree-Fock calculations for H_2, H, F_2 and F; for F_2 the cal-culated dissociation energy has an incorrect sign (see text).

tures from the equilibrium geometry the SCF approach becomes very poor. For the infinite internuclear distance, the SCF energy lies much higher than the double of the SCF energy for the isolated atom. It occurs in most cases that the one-determinant Hartree-Fock function dissociates to incorrect atomic products (giving typically one atom in its excited state). Reasonable dissociation energy for the process $R_2 \longrightarrow 2R$ are not obtained even in the case when they are evaluated directly from the SCF energies for systems R_2 and R. The most striking case is the

F_2 dissociation, for which the Hartree-Fock calculations[256] favor two F atoms with respect to F_2 by 130 kJ/mole. Apparently in dissociations and potential curves involving the electron-pair bond rupture, the correlation energy is of crucial importance and use must be made of some approach which goes beyond the Hartree-Fock limit. A degree of success of some such treatments is discussed in Section 5.D. Sometimes one can arrive at the dissociation energy indirectly. For example, the reaction

$$NaCl \longrightarrow Na + Cl \qquad\qquad (4.12)$$

does not satisfy any of the two conditions for the conservation of correlation energy. A near invariance of the correlation energy may, however, be assumed for the ionic decomposition

$$NaCl \longrightarrow Na^+ + Cl^- \qquad\qquad (4.13)$$

Since NaCl is highly ionic, the two requirements are well satisfied. Since the correlation energies for Na, Cl, Na^+ and Cl^- are safely known, one can arrive in this way[257] at the binding energy of NaCl with chemical accuracy.

To summarize the discussion in this section, it is fair to state that the Hartree-Fock approximation is adequate in many cases. If a rough conservation of the correlation energy may be expected, the Hartree-Fock energy predictions are mostly within 20 kJ/mole or even better. In the past the importance of correlation effects was sometimes overestimated and the discrepancy with experiment found in a particular problem was later attributed to the limited basis set. Very instructive discussions on this problem were reported for the heats of reaction[117,258], the inversion barrier in ammonia[99,118] and the barrier to rotation[99] in H_2O_2.

4.C. Empirical Calculations

Let us start a survey of computational methods by beginning with the simplest conceivable approach[243,259], that of assuming a constant empirical value for each type of bond and therefore obtaining the total correlation energy by summing the bond and lone pair contributions (similarly in the same way that estimates are made in chemistry, e.g., for the heats of formation and dipole moments). We think that this method can hardly be refined to a state that would approach the so--called "chemical accuracy" because it disregards the interpair electron correlation. The data in Table 4.1 show that the interpair con-

T a b l e 4.1
CH-Bond contributions to the correlation energy (results of IEPA calculations[120] performed with localized orbitals; all entries expressed as E/E_h)

Type of contribution	C_2H_2	C_2H_4	C_2H_6	CH_4
Intrapair ε_h	-0.0299	-0.0301	-0.0298	-0.0298
Intergeminal $\varepsilon_{hh'}$		-0.0144	-0.0154	-0.0155
Intervicinal $\varepsilon_{hh'}$	-0.0012	-0.0014	-0.0011	

tributions for neighboring bonds are by no means negligible with respect to intrapair contributions. In the empirical approach the interpair contributions are evidently absorbed in the bond parameters and therefore the effect of the structure cannot be well accounted for.

Empirical parameters are also involved in the EPCE-F2σ method of Sinanoğlu and Pamuk[260,261] as well as the method suggested by Clementi and coworkers[262,263]. The two methods may be called semiempirical because they are based on the formulas given by the theory but adopt approximations involving empirical parameters. In the approach proposed by Clementi and coworkers and also by Colle and Salvetti[264], the formula for the correlation energy is a functional which is a modified expression of Wigner[265] derived for the electron gas. Mathematically, it is an integral containing powers of the electron density. If the Hartree-Fock density is used, the use of the method is restricted to cases where the Hartree-Fock function itself is a good representation of the exact wave function. For example, for dissociation processes the electron density based on a proper (CI) wave function must be employed. Examples of applications in which the use of the Hartree-Fock electron density seems to be sufficient are the water dimer[266] and the complexes[267] of water molecules with Li^+, Na^+, K^+, F^- and Cl^- ions. The correlation energies given by this approach are, however, only semiquantitative and their combination with near Hartree-Fock data cannot be expected to provide highly accurate predictions.

The EPCE-F2σ method of Sinanoğlu and Pamuk[260,261] is based on the Many-Electron Theory (MET) of Sinanoğlu which will be noted in Section 4.G. In its simplest form, MET predicts the correlation energy to be a sum of the pair correlation energies (eqn. (4.47)) which in the LCAO approximation becomes[260]

$$E_{corr} \approx \frac{1}{4} \sum_{\mu} \sum_{\nu} D_{\mu\mu} D_{\nu\nu} \bar{\epsilon}_{\mu\nu} \qquad (4.14)$$

where the indices μ and ν refer to atomic orbitals, the D terms are diagonal elements of the density matrix and

$$\bar{\epsilon}_{\mu\nu} \equiv \epsilon_{\mu^\alpha \nu^\beta} + \epsilon_{\mu^\alpha \nu^\alpha} \qquad (4.15)$$

is the effective pair correlation energy (EPCE) for the orbitals μ and ν for which we assume the "F2 approximation"

$$\epsilon_{\mu^\alpha \nu^\beta} \approx 2\epsilon_{\mu^\alpha \nu^\alpha} \qquad (4.16)$$

The last symbol in the designation of the method means that in contrast to the previous version of the method[268] it is not restricted to π-electronic systems.

Computationally, one performs first a minimum STO basis set or a semiempirical all-valence electron calculation (say CNDO/2) for obtaining electron densities. One-center $\bar{\epsilon}_{\mu\nu}$ are constants which were tabulated[260] for H, B, C, N, O and F atoms. Two-center $\bar{\epsilon}_{\mu\nu}$ are evaluated over STO's by an empirical formula which is a function of the interatomic distance. Computation of the expression (4.14) is much shorter than a standard CNDO/2 run. It is understandable that such a simple method cannot provide highly accurate estimates of the correlation energy. Actually they are claimed[261] to be within ±0.5 eV. Although this tolerance is much larger than that required in many chemical problems, the feasibility of calculations warrants attempts at applications. These were attempted for binding energies[261], ionization potentials[269], excitation energies[269], clusters of hydrogen atoms[270] and intermolecular interactions[271,272].

The presence of the $D_{\mu\mu}$ elements in the formula (4.14) reveals that the same limitations apply to EPCE-F2σ as those noted with the method of Clementi and coworkers. Accordingly, the use of EPCE-F2σ is justifiable only in cases where the SCF approach itself gives a realistic electron density distribution. Moreover, EPCE-F2σ being an approximation to the IEPA method (see Section 4.E.), its use should be restricted to cases where IEPA is known to work well. The two requirements just noted are satisfied with weak intermolecular interaction. Indeed, the attempts at EPCE-F2σ applications in this field were encouraging[272].

A similar idea to that involved in EPCE-F2σ was used[273] for the estimates of correlation energy in different electronic states of diatomic molecules. The essence of that approach is the conversion of the MO representation of the electronic structure of a particular electronic state to the atomic representation by means of the population analysis and the assumption that the predominant correlation effects are due to electron pairs in the same orbital. In this way, the correlation energy is again given by the products of AO populations and AO correlation contributions (available in the literature). Although the approach is rough, it may be useful for spectroscopic purposes. Table 4.2 presents an example of the treatment of this type.

T a b l e 4.2
Term values for the CH radical[274]

State	T_o $(10^3$ cm$^{-1})$		
	SCF	corrected for correlation effects	experimental
A $^2\Delta$	22.7	23.7	23.2
B $^2\Sigma^-$	26.3	26.2	25.6
C $^2\Sigma^+$	33.8	32.0	31.8
D $^2\Pi$	61.0	58.8	59.0

A conclusion that may be drawn on this section is that the empirical methods for estimates of the correlation energy may be profitable in applications and that their further development is worth pursuing. It should be realized that for molecules that are (by their size) of interest to chemists, they mostly represent the only feasible approach.

4.D. Configuration Interaction

We are not going to pay much attention to the CI method because there are numerous reviews on this topic (e.g.,[275-280]). We only briefly note some techniques that treat the CI expansion in a more effective way. We also note the properties of the CI wave function which have some relation to other methods considered in this chapter.

The traditional CI treatment which is nowadays commonly referred to as "brute force" CI is inpractical for routine calculations of correlation energy because of the slow energy convergence of the CI expansion. But for purposes of comparison with other methods it is advantageous to consider just the traditional CI. What we need to know is the role of the singly, doubly, triply, etc. excited configurations in the CI expansion. We limit ourselves to the most common case - the Slater determinant of a closed shell ground state formed from the Hartree-Fock orbitals. In this case, the effect of singly excited configurations is small, as a consequence of Brillouin's theorem. However, it appears that in some special applications (other than energy predictions) single excitations cannot be omitted in the CI expansion. For example, singly excited states are responsible for the correct sign of the dipole moment of CO[281]. The main contribution to the correlation energy, say more than 90%, comes from the doubly excited states. For larger than diatomic molecules, it is difficult, for practical reasons, to test the effect of triple and higher excitations by the way of direct calculations. Therefore it is only possible to do speculative estimates on the basis of data for atoms and diatomic molecules. For example, in the case of the Be atom[282] the triple excitations correspond to 0.3% of the calculated correlation energy and quadruple excitations to 3.8%. A similar percentage was also found[7] with the water molecule: the DZ STO and DZ CGTF basis sets, respectively, gave 1.5 and 0.8% for the triple excitations and 3.0 and 4.3% for the quadruple excitations. In N_2 the contribution[283] of quadruply excited configurations is 7%. It may be assumed, therefore, that for small polyatomic molecules it is sufficient to consider only doubly excited configurations and also to include singly and quadruply excited configurations when striving for highly accurate calculations. With large molecules the situation is less clear-cut. In this respect a pesimistic opinion was reported by Davidson[276] who assumes that for a large molecule, a hydrocarbon chain longer than about 30 carbons, most of the correlation energy should come from the quadruple and higher excitations.

In spite of the truncation of the CI expansion so as to include biexcitations, the ordinary CI calculations are still rather involved. There are many devices and tricks[275-278] which facilitate the problem: use of symmetry, selection of configurations by perturbation estimates of their weights, efficient diagonalization algorithms, economic computer handling of large CI-matrices[284], avoiding the explicit construction of the CI-matrix[285], effective transformations of integrals from AO to MO basis, disregarding configurations that contribute to inner shell correlation energy, truncation of the virtual orbital space[275], energy extrapolations[286] and others. These are, however, of more or less technical nature and do not remove the inherent drawback of the ordinary CI expansion - this being its slow convergence. A very efficient way of obtaining a more rapidly convergent CI expansion is based on the use of natural orbitals[288]. We take note of two techniques of this type. In the Iterative Natural Orbital method (INO-CI) developed by Bender and Davidson[289], one proceeds[99,277] as follows: A certain number of configurations, say 50, is selected and a CI calculation is carried out. The density matrix given by the CI wave function is diagonalized which gives us the natural orbitals. From these the same set of configurations is constructed and the CI calculation is repeated. Several iterations are performed according to this scheme until the energy reaches a minimum. An important feature of the method is that unimportant configurations are deleted in the iteration process and new configurations are added. The choice of singly and doubly excited configurations may be made in an ingenious way (e.g., for H_2O[290] and NH_2[291]) which permits the valence-shell correlation energy to be picked up. Some applications of the method are presented in the book by Schaefer[99]. The second method we are going to note is perhaps more economic. It is the so called Pair Natural Orbital method (PNO-CI) developed by Meyer[292].

As in IEPA-PNO (see Section 4.E.) one calculates the pair natural orbitals for each pair of occupied spinorbitals and constructs from them the (lowest) doubly excited states. But in contrast to IEPA-PNO, the PNO-CI wave functions contain the ground state and doubly excited configurations which correspond to excitations from all pairs. This, of course, brings about difficulties in constructing the Hamiltonian matrix elements, H_{ij}, because each pair generates its own set of PNO's. In other words, PNO's for the pair R, S are not orthogonal to PNO's for the pair T, U. Fortunately the nonorthogonality does not represent a serious problem. Although the CI-matrix H_{ij} elements[293] are somewhat more complex than those given by Slater rules for an orthogonal set, they are still tractable. It should be emphasized that the same H_{ij} elements appear in the CEPA-PNO equations (see Section 4.I.). It is therefore profitable to perform PNO-CI and CEPA-PNO calculations in a single run.

Let us now comment briefly on the general properties of the CI wave function with only singly and doubly excited configurations (referred to as CI-SD approach; it is not of importance whether it was obtained by "brute force", INO-CI or PNO-CI treatment). As we already know it should cover a large portion of the correlation energy. Another advantage is that it furnishes an upper bound to the energy because CI is a genuine variational method. A drawback of the CI-SD wave function is the incorrect dependence on the number of particles (so called size consistency error - see Section 4.M.) and its incorrect behaviour on dissociation (see Section 5.D.). To illustrate the incorrect dependence on the number of particles we make use of the PNO-CI data of Ahlrichs[250] for the dimerization of BH_3. He arrived at the following energies of dimerizations

$$\Delta E = E \ (B_2H_6) - 2E \ (BH_3) = -115 \ kJ/mol \qquad (4.17)$$

$$\Delta E = E \ (B_2H_6) - E \ (2BH_3) = -143 \ kJ/mol \qquad (4.18)$$

In eqn. (4.17) twice the PNO-CI energy of BH_3 was assumed, whereas in eqn. (4.18) the two BH_3 molecules were represented by a supersystem BH_3BH_3 at large intermolecular distance ($R = 50 \ a_o$). In the first case, the two BH_3 molecules are too much favored with respect to B_2H_6. In fact, we consider simultaneous double excitations of either BH_3 molecule, the net result of which means inclusion of quadruply excited configurations. In order to obtain a balanced description of both sides of the reaction, it is necessary to treat it by means of eqn. (4.18). It would be, of course, more rigorous to augment the CI-SD wave functions by higher excitations. Elimination of the size consistency error requires the inclusion of quadruply excited states but, unfortunately, one can hardly do it by selecting a small group of quadruple excitations of a certain type. Hence, the traditional CI approach appears to be of low practical value in attempts at improving the size consistency. One can correct, however, the CI-SD results in an approximate way[287]. Several expressions were suggested that permit to estimate the effect of those quadruple excitations that are responsible for eliminating the incorrect dependence of the CI-SD wave function on the number of particles. The most commonly used expression is due to Langhoff and Davidson[283],

$$\Delta E_Q = (1 - c_0^2) \, \Delta E_D \qquad (4.19)$$

in which ΔE_Q means the correlation energy contribution of quadruple excitations, c_0 is the coefficient of the SCF ground state in the CI wave function generated by the SCF configuration and doubles, and ΔE_D is the contribution of the doubles. An example of the use of eqn. (4.19) is presented in Table 2.17 (column SDQ). The entries show clearly an improved agreement with experiment attained upon applying the correction (4.19). The theoretical grounds of the formula (4.19) were discussed by Bartlett and Shavitt[294]. Other formulas and the discussion on this topic may be found in the cited papers[280,287,295-298]. The physical meaning of eqn. (4.19) may be conveniently demonstrated by means of the perturbation theory. For the sake of compactness, however, we note this problem in Section 4.J.

As regards the multiconfiguration SCF (MCSCF), the virtues and the drawbacks of this method[299-301] have been described in the literature on several occasions (see e.g., Refs 99, 302). In our opinion, MCSCF remains still more suited for treating near degeneracy problems such as removals of discontinuities, cusps, humps and other artefacts in SCF energy hypersurfaces reported by Gregory and Paddon-Row[303,304] rather than to large scale calculations including correlation energy.

4.E. Independent Electron Pair Approximation (IEPA)

This approximation has been known for a long time, e.g., in the form of Sinanoğlu's Many-Electron Theory[305], Nesbet's formulation of the second-order Bethe-Goldstone method[306] or the decoupled equations within the framework of the antisymmetrized product of strongly orthogonal geminals approximation[287]. Practical calculations with the IEPA method, however, were first developed by Kutzelnigg's group[293,307]. In the language of configuration interaction, one uses the wave function

$$\Psi_{AB} = \Phi_0 + \sum_{R<S} c_{AB}^{RS} \Phi_{AB}^{RS} \qquad (4.20)$$

where Φ_0 and Φ_{AB}^{RS}, respectively, are Slater determinants for the ground state and doubly excited configurations. Hereafter spinorbitals will be denoted by capital letters and orbitals by lowercase letters. Equation (4.20) means that we perform a CI calculation, separately, for each pair of occupied spinorbitals A, B. This gives us an energy increment ε_{AB} which can be ascribed to the electron correlation in spinorbitals A, B. Since the electron pairs are considered to be independent, the total correlation energy is given by

$$E_{corr} = \sum_{A<B} \varepsilon_{AB} \qquad (4.21)$$

For technical details of calculations such as the use of pair natural orbitals we refer the reader to the cited papers[293,307]. Here we restrict ourselves to a few comments on the properties of the IEPA correlation energy. IEPA is not a variational method and it does not give an upper bound to the energy. Actually, the correlation energy is overestimated, in some cases by as much as 30%. The extent of this overestimation depends on the particular case and it is, therefore, possible to speak about "good" and "poor" IEPA molecules[120]. An example of the failure of IEPA is presented for the F_2 dissociation in Section 5.D. Among the advantages of IEPA, the major one is the computational economy. This allows one to undertake systematic studies of the effect of correlation energy on molecular geometries and force constants[118,308-311], potential curves[120,311], and van der Waals interactions[312-316]. Another advantage of IEPA is the correct dependence[240,293] of the correlation energy on the number of electrons, i.e. the energy of a supersystem of n noninteracting subsystems, E(nA), is equal to the sum of the energies of the subsystems, nE(A) (see Section 4.M.).

4.F. Cluster Expansion of the Wave Function

In this section we shall discuss an approach which is neither variational nor perturbational. This approach has its origin in nuclear physics and was introduced to quantum chemistry by Sinanoğlu[305]. It is based on a cluster expansion of the wave function. A systematic method for calculation of cluster expansion components of the exact wave function was developed by Čížek[317]. The characteristic feature of this approach is the expansion of the wave function as a linear combination of Slater determinants. Formally, this expansion is similar to the ordinary CI expansion. The cluster expansion, however, gives us not only physical insight into the correlation energy but it also shows the connections between the variational approaches (CI) and the perturbational approaches.

We shall express the exact wave function in the form

$$\Psi = \Phi_o + \eta \qquad (4.22)$$

where Φ_o is the Slater determinant for the closed shell ground state and η is the correlation function, which describes the correlation of two, three etc. electrons. We can choose the correlation function for a N-electron system in the form

$$\eta = \sum_i u^{(i)} + \sum_{i<j} u^{(ij)} + \sum_{i<j<k} u^{(ijk)} + \sum_{i<j<k<\ell} u^{(ijk\ell)} + .. \quad (4.23)$$

where the indices $i, j, k, \ell, \ldots$ are indices of the electrons 1 to N. The functions $u^{(i)}$ we call one-electron clusters, $u^{(ij)}$ two-electron clusters, $u^{(ijk)}$ three-electron clusters, etc. Let us examine in more detail the term $u^{(ijk\ell)}$ which describes the correlation of four electrons. We can distinguish the simultaneous correlation of four electrons (so called linked cluster), for example, from the interaction of two pairs of electrons (so called unlinked cluster). Although commonly used, the term "linked cluster" is not very fortunate because it is in conflict with other meanings. For this reason use is sometimes made of the term "connected cluster". Simultaneous four electron correlation occurs only in situations when all four electrons are close together. Since such "collisions" are rare in molecules we can expect that the effect of linked four electron clusters in expansion (4.23) will not be important[305]. On the other hand, the effect of unlinked clusters may be important since these correspond to a collision of two electrons i and j and independent and simultaneous collision of another two electrons k and ℓ. Obviously, collisions of another type are conceivable and any cluster can therefore be described as the sum of linked and unlinked clusters

$$U^{(i)} = t^{(i)} \Phi_o$$

$$U^{(ij)} = t^{(ij)} \Phi_o + t^{(i)} t^{(j)} \Phi_o$$

$$\tag{4.24}$$

$$U^{(ijk)} = t^{(ijk)} \Phi_o + t^{(k)} t^{(ij)} \Phi_o + t^{(j)} t^{(ik)} \Phi_o$$

$$+ t^{(i)} t^{(jk)} \Phi_o + t^{(i)} t^{(j)} t^{(k)} \Phi_o$$

etc.

In equations (4.24) the t operators are generating the clusters U from the Slater determinant Φ_o and their effect on Φ_o may be viewed as follows

$$\sum_i t^{(i)} \Phi_o = \sum_A \sum_R d_A^R \Phi_A^R \tag{4.25}$$

$$\sum_{i<j} t^{(ij)} \Phi_o = \sum_{A<B} \sum_{R<S} d_{AB}^{RS} \Phi_{AB}^{RS} \tag{4.26}$$

$$\sum_{i<j<k} t^{(ijk)} \Phi_o = \sum_{A<B<C} \sum_{R<S<T} d_{ABC}^{RST} \Phi_{ABC}^{RST} \tag{4.27}$$

Let us define

$$T_1 = \sum_i t^{(i)} \tag{4.28}$$

$$T_2 = \sum_{i<j} t^{(ij)} \tag{4.29}$$

$$T_3 = \sum_{i<j<k} t^{(ijk)} \tag{4.30}$$

and

$$T = T_1 + T_2 + \ldots T_N \tag{4.31}$$

where N means the number of electrons. By introducing eqns. (4.23)-
(4.31) to eqn. (4.22), the wave function may be expressed in a very
compact form (for the references to the original papers and the theoretical background see cited reviews[318,319])

$$\Psi = e^T \Phi_o \tag{4.32}$$

because expanding e^T gives us

$$\Psi = \left(1 + T + \frac{T^2}{2!} + \frac{T^3}{3!} + \ldots\right) \Phi_o \tag{4.33}$$

Let us compare this cluster expansion with the well known CI expansion
of the wave function

$$\Psi = c_o \Phi_o + \sum_A \sum_R c_A^R \Phi_A^R + \sum_{A<B} \sum_{R<S} c_{AB}^{RS} \Phi_{AB}^{RS} + \ldots \tag{4.34}$$

We see that the cluster expansion is formally the same, only instead
of the c-set of expansion coefficients we have the d-coefficients
(appearing in eqns. (4.25)-(4.27)). Comparing coefficients standing
before respective configurations gives us the following relations[319]

$$c_o^{-1} c_A^R = d_A^R \tag{4.35}$$

$$c_o^{-1} c_{AB}^{RS} = d_{AB}^{RS} + d_A^R d_B^S - d_A^S d_B^R \tag{4.36}$$

$$c_o^{-1} c_{ABC}^{RST} = d_{ABC}^{RST} + d_A^R d_{BC}^{ST} - d_B^R d_{AC}^{ST} + d_C^R d_{AB}^{ST}$$

$$- d_A^S d_{BC}^{RT} + d_B^S d_{AC}^{RT} - d_C^S d_{AB}^{RT} + d_A^T d_{BC}^{RS}$$

$$- d_B^T d_{AC}^{RS} + d_C^T d_{AB}^{RS} + d_A^R d_B^S d_C^T$$

$$- d_A^R d_B^T d_C^S - d_A^S d_B^R d_C^T + d_A^S d_B^T d_C^R + d_A^T d_B^R d_C^S - d_A^T d_B^S d_C^R \qquad (4.37)$$

A question may now be asked why we are attempting an expansion of the form (4.32) which is actually more complex than the ordinary CI expansion. We shall show that this expansion can be considerably reduced without losing much rigor. Before doing so, however, it is profitable to examine first the relative importance of individual types of clusters. We have already noted the clusters $t^{(ijk\ell)} \Phi_o$ and $t^{(ij)} t^{(k\ell)} \Phi_o$. As regards the clusters $t^{(i)} \Phi_o$, their contribution to the correlation energy will be small when Hartree-Fock orbitals are used. It is now also understandable that all products containing $t^{(i)}$ such as $t^{(i)} t^{(j)}$, $t^{(i)} t^{(jk)}$ etc. will be small. The effect of linked tri-excited clusters is hardly to be assessed. The calculation which will be discussed in Section 4.H. suggests that it is small, but the paucity of numerical results reported in the literature precludes any generalization. It is only possible to state that from three-electron clusters upwards the contribution to the correlation energy coming from linked clusters decreases rapidly. From the above discussion it is possible to conclude that the most important clusters are $T_2 \Phi_o$ and $1/2\ T_2^2 \Phi_o$. This finding is equivalent to assuming $T \approx T_2$ in eqn. (4.32).

We shall now draw our attention to the practical use of the formalism of the cluster expansion. Our goal is to solve again the equation

$$H |\Psi\rangle = E |\Psi\rangle \qquad (4.38)$$

where $|\Psi\rangle$ is the exact wave function and E is the exact energy of a N-electron system. Let us substitute the expansion (4.32) into eqn. (4.38) which gives us

$$H\ e^T |\Phi_o\rangle = E\ e^T |\Phi_o\rangle \qquad (4.39)$$

From the arguments given above it appears that it is reasonable to assume $T \approx T_2$. Equation (4.39) becomes

$$H e^{T_2} |\Phi_0\rangle = E e^{T_2} |\Phi_0\rangle \qquad (4.40)$$

or

$$H \left\{ 1 + T_2 + \frac{T_2^2}{2!} + \ldots \right\} |\Phi_0\rangle = E \left\{ 1 + T_2 + \frac{T_2^2}{2!} + \ldots \right\} |\Phi_0\rangle \qquad (4.41)$$

Let us subtract the term $\langle \Phi_0 | H | \Phi_0 \rangle$ from both sides of eqn. (4.41). This gives us

$$\left\{ H - \langle \Phi_0 | H | \Phi_0 \rangle \right\} \left\{ 1 + T_2 + \frac{T_2^2}{2} + \ldots \right\} |\Phi_0\rangle$$

$$= \left\{ E - \langle \Phi_0 | H | \Phi_0 \rangle \right\} \left\{ 1 + T_2 + \frac{T_2^2}{2} + \ldots \right\} |\Phi_0\rangle \qquad (4.42)$$

The term $\left\{ E - \langle \Phi_0 | H | \Phi_0 \rangle \right\}$ is the correlation energy of the closed shell ground state.

In principle, the problem (4.42) may be solved in the traditional way. However, it is much more advantageous to use what is referred to as a "many-body" approach. This means expressing eqn. (4.42) in the second quantization formalism, applying Wick's theorem, constructing pertinent Feynman-like diagrams and assigning to diagrams the final mathematical expressions according to certain rules. The whole technique is nicely described in the review by Paldus and Čížek[320]. It should be emphasized that for a more general cluster expansion containing also other than T_2 and T_2^2 clusters, the problem is practically tractable only by making use of the diagrammatic approach.

A rigorous solution of the problem (4.42) leads to the CPMET equations[317] (see Section 4.H.). However, prior to discussing the CPMET method, we note in the following section the pioneering work of Sinanoğlu.

4.G. Many-Electron Theory (MET) of Sinanoğlu

Sinanoğlu was the first who suggested[305] a practical method for calculating the correlation energy based on the cluster expansion of the wave function. By the approximate treatment of the problem (4.42) he arrived for the function

$$\Psi = \Phi_o + \sum_{i<j} t^{(ij)}\, \Phi_o \tag{4.43}$$

at the following expression

$$E \approx E_{HF} + \frac{1}{D} \sum_{i<j} \tilde{\varepsilon}_{ij} \tag{4.44}$$

in which $\tilde{\varepsilon}_{ij}$ are pair correlation energies and D is related to the normalization of Ψ. Pair correlations are determined from the so called pair functions (in our notation these are clusters $t^{(ij)}\,\Phi_o$) which are obtained independently for each pair of occupied spinorbitals. In this simplest form, MET is equivalent to IEPA (see Section 4.E.). If the unlinked clusters are included in an approximation, Sinanoğlu can transform the function

$$\Psi = \Phi_o + \sum_{i<j} t^{(ij)}\, \Phi_o + \sum_{\substack{i<j\ k<\ell \\ i,j \neq k,\ell}} \sum t^{(ij)} t^{(k\ell)}\, \Phi_o \tag{4.45}$$

into the following expression

$$E \approx E_{HF} + \sum_{i<j} \tilde{\varepsilon}_{ij}\, \frac{D_{ij}}{D'} \tag{4.46}$$

This equation and (4.44) differ only by normalization factors. For $N \to \infty$ it was shown[305] that $(D_{ij}/D') \to 1$, so that it is possible to write

$$E \approx E_{HF} + \sum_{i<j} \tilde{\varepsilon}_{ij} \tag{4.47}$$

The most important result of the papers by Sinanoğlu on MET was the finding that in a CI treatment with quadruply excited configurations, the linked tetra-excited part (T_4) is negligible in comparison to the unlinked part ($1/2\ T_2^2$). This was clearly shown by Sinanoğlu in an a- nalysis of the CI calculation performed by Watson[321] for the beryllium atom. We present Sinanoğlu's analysis[305] in a somewhat modified way in order to be consistent with the definitions of c and d expansion coef- ficients (in Section 4.F.). Watson's CI wave function contained 37 most important configurations. Among them were four quadruply excited configurations. These are configurations 5, 7, 11 and 13 of Table 4.3. The other configurations listed in Table 4.3 are doubly excited con- figurations relevant for expressing the respective unlinked clusters. As we shall learn in the next section, the "unlinked part" of the ex- pansion coefficient for a quadruply excited configuration may be ex- pressed by products of expansion coefficients for doubly excited states (eqns. (4.53) and (4.51)). The C_7 coefficient, for example, may accordingly be approximated as

$$c_7 \approx c_0 \left[c_1 c_3 + 2 \left(\frac{1}{\sqrt{2}} c_8 \right) \left(\frac{1}{\sqrt{2}} c_{12} \right) \right] = 0.00630 \tag{4.48}$$

The coefficients c_5, c_7, c_{11} and c_{13} are presented in Table 4.4. One can see that the data given by the wave function (4.45) are in good agreement with the results of the complete CI treatment and that the effect of linked T_4 clusters is therefore very small.

Sinanoğlu also derived the LCAO form of MET and suggested a series of semiempirical procedures for estimating the correlation energy. Among them the one of most general use is the so called EPCE-F2σ meth- od formulated by Sinanoğlu and Pamuk[260,261] (see Section 4.C.).

4.H. Coupled-Pair Many-Electron Theory (CPMET)

We present here a simple derivation[319,322] of CPMET equations in a traditional way. On applying the restriction $T \approx T_2$, our starting point becomes

T a b l e 4.3

Selected configurations of the CI wave function for the beryllium atom[321]

Configuration[a] (Φ_n)	Label n	Energy contribution of Φ_n, $(E_n - E_{n-1})/E_h$	CI expansion coefficient, c_n
$1s^2 2s^2$	0	-14.57299	0.9575824
$1s^2 p_I^2$	1	-0.04116	-0.2844586
$2s^2 p_{II}^2$	2	-0.01769	-0.0262111
$2s^2 s_I^2$	3	-0.01071	-0.0232595
$2s^2 d_{II}^2$	4	-0.00213	-0.0059003
$p_I^2 p_{II}^2$	5	-0.00157	0.0070633
$2s^2 p_I p_{II}$	6	-0.00106	-0.0073734
$p_I^2 s_I^2$	7	-0.00100	0.0056508
$1s2s s_I^2$	8	-0.00084	-0.0055013
$1s2s p_{II}^2$	9	-0.00063	-0.0047417
$1s^2 d_I^2$	10	-0.00040	-0.0182294
$p_I^2 d_{II}^2$	11	-0.00019	0.0015848
$1s2s p_I^2$	12	-0.00020	0.0065906
$p_{II}^2 d_I^2$	13	-0.00000	0.0004639
$2s^2 p_I^2$	14	-0.00003	-0.0018990
$1s^2 p_I p_{II}$	15	-0.00009	0.0025567
$1s^2 p_{II}^2$	16	-0.00000	-0.0003084

[a] Roman numerals denote the order of virtual orbitals.

T a b l e 4.4

Four-electron correlation and unlinked clusters in Be atom

Quadruply excited configuration[a]	Coefficient from 37-configuration wave function[321]	Coefficient calculated from double excitations by eqns.(4.51) and (4.53)
$p_I^2 p_{II}^2$	0.007063	0.0071
$p_I^2 s_I^2$	0.005651	0.00630
$p_I^2 d_{II}^2$	0.001585	0.00161
$p_{II}^2 d_I^2$	0.000464	0.000458
Energy contribution[305]	-0.075 eV	-0.074 eV

[a] Roman numerals denote the order of virtual orbitals.

$$H \, e^{T_2} | \Phi_o \rangle \; = E \, e^{T_2} | \Phi_o \rangle \tag{4.49}$$

and the comparison of cluster and CI expansions represented by eqns. (4.35)-(4.37) reduces to

$$c_A^R = 0 \tag{4.50}$$

$$c_o^{-1} c_{AB}^{RS} = d_{AB}^{RS} \tag{4.51}$$

$$c_{ABC}^{RST} = 0 \tag{4.52}$$

$$
\begin{aligned}
c_o^{-1} c_{ABCD}^{RSTU} = \; & d_{AB}^{RS} d_{CD}^{TU} - d_{AC}^{RS} d_{BD}^{TU} + d_{AD}^{RS} d_{BC}^{TU} - d_{AB}^{RT} d_{CD}^{SU} + d_{AC}^{RT} d_{BD}^{SU} \\
& - d_{AD}^{RT} d_{BC}^{SU} + d_{AB}^{RU} d_{CD}^{ST} - d_{AC}^{RU} d_{BD}^{ST} + d_{AD}^{RU} d_{BC}^{ST} + d_{AB}^{TU} d_{CD}^{RS} \\
& - d_{AC}^{TU} d_{BD}^{RS} + d_{AD}^{TU} d_{BC}^{RS} - d_{AB}^{SU} d_{CD}^{RT} + d_{AC}^{SU} d_{BD}^{RT} - d_{AD}^{SU} d_{BC}^{RT} \\
& + d_{AB}^{ST} d_{CD}^{RU} - d_{AC}^{ST} d_{BD}^{RU} + d_{AD}^{ST} d_{BC}^{RU} \tag{4.53}
\end{aligned}
$$

We now project eqn. (4.49) into the spaces spanned by $| \Phi_o \rangle$ and doubly excited Slater determinants $| \Phi_{AB}^{RS} \rangle$, respectively. This gives us

$$\langle \Phi_o | H \, e^{T_2} | \Phi_o \rangle \; = \langle \Phi_o | H(1 + T_2) | \Phi_o \rangle \; = E \langle \Phi_o | e^{T_2} \Phi_o \rangle \; = E \tag{4.54}$$

and

$$
\begin{aligned}
\langle \Phi_{AB}^{RS} | H \, e^{T_2} | \Phi_o \rangle \; &= \langle \Phi_{AB}^{RS} | H(1 + T_2 + \frac{T_2^2}{2}) | \Phi_o \rangle \\
&= E \langle \Phi_{AB}^{RS} | e^{T_2} \Phi_o \rangle \; = E \langle \Phi_{AB}^{RS} | (1 + T_2) \Phi_o \rangle \; = E \, d_{AB}^{RS} \tag{4.55}
\end{aligned}
$$

Substituting for E in eqn. (4.55) by (4.54) we get

$$\left\langle \Phi_{AB}^{RS} \middle| H\left(1 + T_2 + \frac{1}{2}T_2^2\right) \middle| \Phi_o \right\rangle = \left\langle \Phi_o \middle| H(1 + T_2) \middle| \Phi_o \right\rangle d_{AB}^{RS} \qquad (4.56)$$

which expressed in terms of matrix elements leads to CPMET equations[317]

$$\left\langle \Phi_{AB}^{RS} \middle| H \middle| \Phi_o \right\rangle + \sum_{C<D} \sum_{T<U} \left\langle \Phi_{AB}^{RS} \middle| H \middle| \Phi_{CD}^{TU} \right\rangle d_{CD}^{TU}$$

$$+ \frac{1}{2} \sum_{\substack{C<D \\ C,D \neq A,B}} \sum_{\substack{T<U \\ T,U \neq R,S}} \left\{ \left(d_{AB}^{RS} d_{CD}^{TU} - d_{AC}^{RS} d_{BD}^{TU} + \cdots \right) + \left(d_{CD}^{TU} d_{AB}^{RS} \right. \right.$$

$$\left. \left. - d_{BD}^{TU} d_{AC}^{RS} + \cdots \right) \right\} \left\langle \Phi_{AB}^{RS} \middle| H \middle| \Phi_{ABCD}^{RSTU} \right\rangle = \left\langle \Phi_o \middle| H \middle| \Phi_o \right\rangle d_{AB}^{RS}$$

$$+ \sum_{C<D} \sum_{T<U} \left\langle \Phi_o \middle| H \middle| \Phi_{CD}^{TU} \right\rangle d_{CD}^{TU} d_{AB}^{RS} \qquad (4.57)$$

The advantage of the many-body approach[317] over that represented by eqns. (4.49)-(4.57) is that it yields the CPMET equations directly in terms over orbitals instead of expressions containing CI matrix elements. Moreover, the many-body approach is quite general and permits arbitrary clusters to be included in the e^T expansion. Hence the CPMET represents an outstanding tool for examining rigorously effects of different clusters. Unfortunately, up to now only a few applications of that kind have been reported[323-326]. We comment on the first of them, that one due to Paldus, Čížek and Shavitt[323], which aimed at the comparison of CPMET with the full CI (FCI) calculation of Pipano and Shavitt[327] for BH_3. In that study CPMET was extended to account also for T_1, $T_1 T_2$ and T_3 clusters. This extended version of CPMET is called[323] ECPMET. Testing of ECPMET was performed[323] with the same basis set as the FCI calculation of Pipano and Shavitt. It was shown that the effect of linked four electron and higher clusters (both unlinked and linked) corresponds to 0.002% of the total E_{corr} given by this basis set. The effect of T_1 clusters is very small (less than

0.1% of the total computed correlation energy). The approximation $T \approx T_2$ ($e^T = 1 + T_2 + 1/2 \, T_2^2$) gives a correlation energy almost identical to that obtained by CI-DQ (configuration interaction for the ground state, doubly and quadruply excited configurations) in which linked and unlinked contributions are lumped together. (The contribution of neglected linked clusters amounts to 0.004%.) Also the effect of relative importance of linked (T_3) and unlinked ($T_1 T_2$) clusters was tested. It was concluded that, as far as the calculation of the correlation energy is concerned, the relative importance of linked and unlinked terms for tri-excited clusters is just the opposite of that found for tetra-excited clusters: the role of unlinked tri-excited clusters is negligible compared to the linked tri-excited terms. Thus, the contribution of tri-excited clusters is predominantly due to linked terms. The overall contribution of tri-excited clusters was computed to be less than 0.8% of the total correlation energy.

As regards computational aspects, in CPMET we have to solve a system of equations (4.57) which expressed in matrix elements over orbitals has the following general form[317,328]

$$\sum_j a_{ij} x_j + \sum_{j<k} b_{ijk} x_j x_k + c_i = 0 \qquad (4.58)$$

where x_i stands for the unknown d coefficients of the cluster expansion that are to be determined, c_i are two-electron repulsion integrals, and a_{ij} and b_{ijk} are coefficients containing matrix elements of the Hartree-Fock and $1/r_{12}$ operators. The indices run over all distinct doubly excited configurations. The system of equations (4.58) can we solved iteratively.

CPMET is obviously a nonvariational method. However, the advantage of having an upper bound to the energy is probably not so important when the method is accurate enough to give the correlation energy with an accuracy of a few percent.

4.I. Coupled-Electron Pair Approximation (CEPA)

CEPA[292,329] represents one of the most successful approaches to the calculation of correlation energy of molecules from the viewpoint of accuracy and the expense of computer line. Its formulation was prompted by a rather complex form of the CPMET equations. It was hoped that the rigor of CPMET might be sacrificed for gaining much in the computational effort but losing little in accuracy. The approximation adopted in CEPA refers to the third term on the left hand side of CPMET equations (4.57) for which it is assumed

$$\frac{1}{2} \sum_{\substack{C<D \\ C,D \neq A,B}} \sum_{\substack{T<U \\ T,U \neq R,S}} \left\{ \left(d_{AB}^{RS} d_{CD}^{TU} - d_{AC}^{RS} d_{BD}^{TU} + \ldots \right) + \left(d_{CD}^{TU} d_{AB}^{RS} \right. \right.$$

$$\left. \left. - d_{BD}^{TU} d_{AC}^{RS} + \ldots \right) \right\} \left\langle \Phi_{AB}^{RS} | H | \Phi_{ABCD}^{RSTU} \right\rangle \approx d_{AB}^{RS} \sum_{\substack{C,D \neq A,B}} \sum_{T<U}$$

$$\left\langle \Phi_{CD}^{TU} | H | \Phi_o \right\rangle d_{CD}^{TU} \tag{4.59}$$

Note that the terms in parentheses on the left hand side of eqn. (4.59) contain products of d coefficients and represent the unlinked part of the CI expansion coefficient for the quadruply excited configuration Φ_{ABCD}^{RSTU} (compare eqn. (4.53)). Within the CEPA these terms are now substituted by only a single product of d's, which may be assumed to have a dominant effect (to see it numerically, substitute for c-coefficients in eqn. (4.48) with values given in Table 4.3). Hence, as in CPMET, the wave function is assumed in the form

$$\Psi = e^{T_2} \Phi_o \tag{4.60}$$

but in contrast to CPMET, the unlinked $T_2 T_2$ clusters are treated in an approximate way. This approximation was suggested by Kelly[330,331] but was first employed for practical calculations by Meyer. The effect of the neglected terms in eqn. (4.59) may be estimated by comparing the results of CEPA calculations with the data given by, for example, Čížek's CPMET (see Section 4.H) or the perturbation theory through fourth order (see Section 4.J.), in which the $T_2 T_2$ clusters are treated rigorously. Here we only state qualitatively[16,322] that the approximation (4.59) should work well if electron pairs are well separated and if also the corresponding pairs of virtual spinorbitals are well localized.

Introducing the approximation (4.59) into the CPMET equations

(4.57) we obtain

$$\langle \Phi_{AB}^{RS} |H| \Phi_o \rangle + \sum_{C<D} \sum_{T<U} \langle \Phi_{AB}^{RS} |H| \Phi_{CD}^{TU} \rangle \; d_{CD}^{TU} +$$

$$+ \; d_{AB}^{RS} \sum_{C,D \neq A,B} \sum_{T<U} \langle \Phi_{CD}^{TU} |H| \Phi_o \rangle \; d_{CD}^{TU}$$

$$= \langle \Phi_o |H| \Phi_o \rangle \; d_{AB}^{RS} + \sum_{C<D} \sum_{T<U} \langle \Phi_o |H| \Phi_{CD}^{TU} \rangle \; d_{CD}^{TU} d_{AB}^{RS} \qquad (4.61)$$

From inspection of eqn. (4.54) it follows that

$$E - \langle \Phi_o |H| \Phi_o \rangle = \sum_{C<D} \sum_{T<U} \langle \Phi_o |H| \Phi_{CD}^{TU} \rangle \; d_{CD}^{TU} \qquad (4.62)$$

We can now define the pair correlation energy ε_{CD} as

$$\varepsilon_{CD} = \sum_{T<U} \langle \Phi_o |H| \Phi_{CD}^{TU} \rangle \; d_{CD}^{TU} \qquad (4.63)$$

which permits us to obtain the CEPA equations in a very compact form

$$\langle \Phi_{AB}^{RS} |H| \Phi_o \rangle + \sum_{C<D} \sum_{T<U} \langle \Phi_{AB}^{RS} |H| \Phi_{CD}^{TU} \rangle \; d_{CD}^{TU}$$

$$= d_{AB}^{RS} \langle \Phi_o |H| \Phi_o \rangle + d_{AB}^{RS} \varepsilon_{AB} \qquad (4.64)$$

The CEPA equations (4.64) are solved iteratively. One may start with the IEPA pair correlation energies, for example, and obtain the d coefficients from eqn. (4.64). These can then be used to evaluate new pair correlation energies using eqn. (4.63) which then can be used in eqn. (4.64) in the next step.

The CEPA computer programs, developed by two German groups[293,329] are based on the use of pair natural orbitals[288] and, therefore, it is appropriate to refer to their approach as the CEPA-PNO method. The construction of H-matrix elements over the PNO's is the same as with PNO-CI (see Section 4.D.).

CEPA was tested systematically on a series of small molecules[118-120,332-334]. Excellent agreement with experiment was found for molecular geometries, spectroscopic constants, dipole moments, dissociation energies, ionization potentials and electron affinities. As regards the potential curves, CEPA gives very good agreement with experiment over a relatively large region around the equilibrium distance. At larger distances the CEPA energy starts to deviate from the experimental curve and it does not converge towards the correct dissociation limit (for details see Section 5.D.).

CEPA is not a variational method, but the high accuracy achieved in the calculated properties suggests that the advantage of furnishing an upper bound to the energy is no so important. In contrast to PNO-CI, the approximate inclusion of unlinked clusters ensures the correct dependence with respect to the number of particles. As with any theoretical approach going beyond the Hartree-Fock level, the portion of the correlation energy accounted for depends on the size of the basis set. Polarization functions were found to be very important because they contribute much more to the correlation energy than to the SCF energy [118,119] (see Section 4.L.). Roughly speaking, CEPA gives about 85% of the total correlation energy for basis sets containing two sets of polarization functions[118]. To illustrate the computer time required for CEPA calculations we present in Table 4.5 the data reported by Ahlrichs et al.[118]

4.J. Perturbation Calculations

This section is devoted to a very perspective approach which is based on the Many-Body Rayleigh-Schrödinger Perturbation Theory (MB-RSPT). What is commonly referred to as MB-RSPT is developed by second quantization and Wick's theorem which are used to give the diagrammatic description of ordinary time-independent Rayleigh-Schrödinger perturbation theory. The use of the term "many-body", which originates from nuclear physics, is justifiable because the explicit expressions in MBPT are expressed in terms of matrix elements of spin-orbitals or orbitals which reflect the many-electron interaction. It is an advantage of using Feynman-like diagrams, which gives a "microscopic" view to the electron interaction in atoms and molecules.

The whole theoretical background is described in several recent

T a b l e 4.5

CEPA calculations[118] of planar NH_3 with 172 Gaussian lobes in 58 groups on a UNIVAC 1108

Operation[a]	CPU time (in min.)
Evaluation of the integrals	73
Hartree-Fock (7 iterations)	4
Construction of operators[b]	12
Calculation of PNO (11 pairs)[c]	12
Construction of the diagonal CI blocks	76
Construction of the off-diagonal blocks	4
Evaluation of E_{CEPA} and E_{CI}[d]	2
Miscellaneous	4
Total	187

[a] For details see Ref. 293.
[b] Coulomb J^R, J^{RS} and exchange K^R and K^{RS} operators.
[c] The other five pairs are obtained by reflection.
[d] CEPA-PNO and PNO-CI calculations are closely related and they are performed in a single run.

reviews[238,319,320,335-337]. We restrict ourselves here to results that were obtained with the Møller-Plesset[338] partitioning of the Hamiltonian, which means that the Hartree-Fock operator was extracted from the Hamiltonian as the "unperturbed" operator and the rest of the Hamiltonian was taken as the perturbation. The formula for the correlation energy of the closed shell ground state may be expressed in a very compact form[320,339]. We only outline here main features of what has been derived in detail elsewhere[320]. Let us assume that a perturbed Hamiltonian of an atomic or molecular system, K, may be split as

$$K = K_0 + W \qquad (4.65)$$

where K_0 is the unperturbed Hamiltonian and W is the perturbation. In order to obtain a direct expression for the correlation energy we use the notation K for our Hamiltonian. As will later be seen K differs from the usual Hamiltonian by a scalar quantity. We assume that the following equations hold for K and K_0 operators

$$K | \Psi_i \rangle = E_i | \Psi_i \rangle \tag{4.66}$$

$$K_0 | \Phi_i \rangle = \varkappa_i | \Phi_i \rangle \tag{4.67}$$

Equation (4.67) represents actually the Hartree-Fock problem and we assume that its complete solution is known. Our goal is to find the solution of eqn. (4.66) under the assumption that Φ changes into Ψ if the perturbation W is switched on. RSPT gives us

$$E_i = \varkappa_i + \sum_{n=0}^{\infty} \langle \Phi_i | W [Q_i (W + \varkappa_i - E_i)]^n | \Phi_i \rangle \tag{4.68}$$

where

$$Q_i = \sum_{\substack{j \\ (j \neq i)}} \frac{| \Phi_j \rangle \langle \Phi_j |}{\varkappa_i - \varkappa_j} = \frac{1 - | \Phi_i \rangle \langle \Phi_i |}{\varkappa_i - K_0} \tag{4.69}$$

Equation (4.68) can be solved iteratively. We can collect the terms having the same order of perturbation and therefore can write

$$E_i = \sum_{j=0} E_i^{(j)} \tag{4.70}$$

where $E_i^{(j)}$ is the j-th contribution. The terms up to the third order have the following forms

$$E_i^{(0)} = \langle \Phi_i | K_0 | \Phi_i \rangle \tag{4.71}$$

$$E_i^{(1)} = \langle \Phi_i | W | \Phi_i \rangle \tag{4.72}$$

$$E_i^{(2)} = \langle \Phi_i | W Q_i W | \Phi_i \rangle \tag{4.73}$$

$$E_i^{(3)} = \langle \Phi_i | W Q_i (W - E_i^{(1)}) Q_i W | \Phi_i \rangle \tag{4.74}$$

The whole problem of calculating E_i (at least up to the third order) is now reduced to the calculation of individual terms (4.71)-(4.74). It is profitable to specify the operators K and K_0 as follows

$$K = H - \langle \Phi_0 | H | \Phi_0 \rangle \tag{4.75}$$

and

$$K_0 = H_0 - \langle \Phi_0 | H_0 | \Phi_0 \rangle \tag{4.76}$$

where H and H_0, respectively, have the usual meaning of the Hamiltonian and the Hartree-Fock operator

$$H | \Psi_i \rangle = \mathcal{E}_i | \Psi_i \rangle \tag{4.77}$$

$$H_0 | \Phi_i \rangle = e_i | \Phi_i \rangle \tag{4.78}$$

Here $\mathcal{E}_i$ is the exact total energy of the system, Φ_i are solutions of the Hartree-Fock problem, and e_i is the sum of Hartree-Fock orbital energies over occupied spinorbitals. Then the eigenvalue in eqn. (4.66), E_i, becomes directly the correlation energy in the i-th electronic state. Since our concern is focused on the ground state, i.e. i = 0, the index i in eqn. (4.70) may be dropped and the respective contributions to the correlation energy can be expressed as

$$E^{(0)} = 0 \tag{4.79}$$

$$E^{(1)} = 0 \tag{4.80}$$

$$E^{(2)} = \langle \Phi_0 | W Q_0 W | \Phi_0 \rangle \tag{4.81}$$

$$E^{(3)} = \langle \Phi_0 | W Q_0 W Q_0 W | \Phi_0 \rangle \tag{4.82}$$

$$E^{(4)} = \langle \Phi_0 | WQ_0 WQ_0 WQ_0 W | \Phi_0 \rangle - \langle \Phi_0 | WQ_0 W | \Phi_0 \rangle \langle \Phi_0 | WQ_0 Q_0 W | \Phi_0 \rangle$$

$$(4.83)$$

For the purposes of the forthcoming discussion of correlation contributions through $E^{(4)}$, it is profitable to introduce the diagrammatical
representation. Let us start with the second and third order contributions, $E^{(2)}$ and $E^{(3)}$, for which the corresponding diagrams are presented in Figs. 4.3 and 4.4. It may be noticed that these figures con-

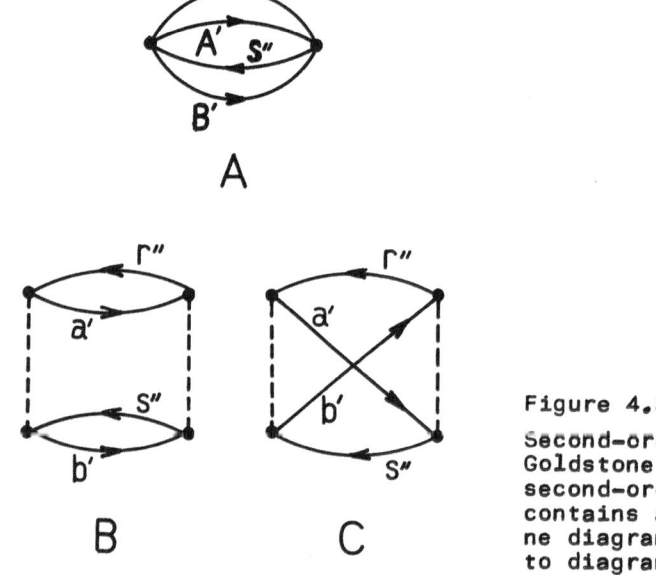

Figure 4.3

Second-order Hugenholtz (A) and
Goldstone (B, C) diagrams. The
second-order contribution, $E^{(2)}$,
contains also two other Goldstone diagrams that are equivalent
to diagrams B and C.

tain two types of diagrams: Hugenholtz - type diagrams (A in Fig. 4.3)
and Goldstone - type diagrams (B and C in Fig. 4.3). The relation between Hugenholtz and Goldstone diagrams is given by

$$(4.84)$$

Each diagram in Figs. 4.3 and 4.4 represents a summation over in-

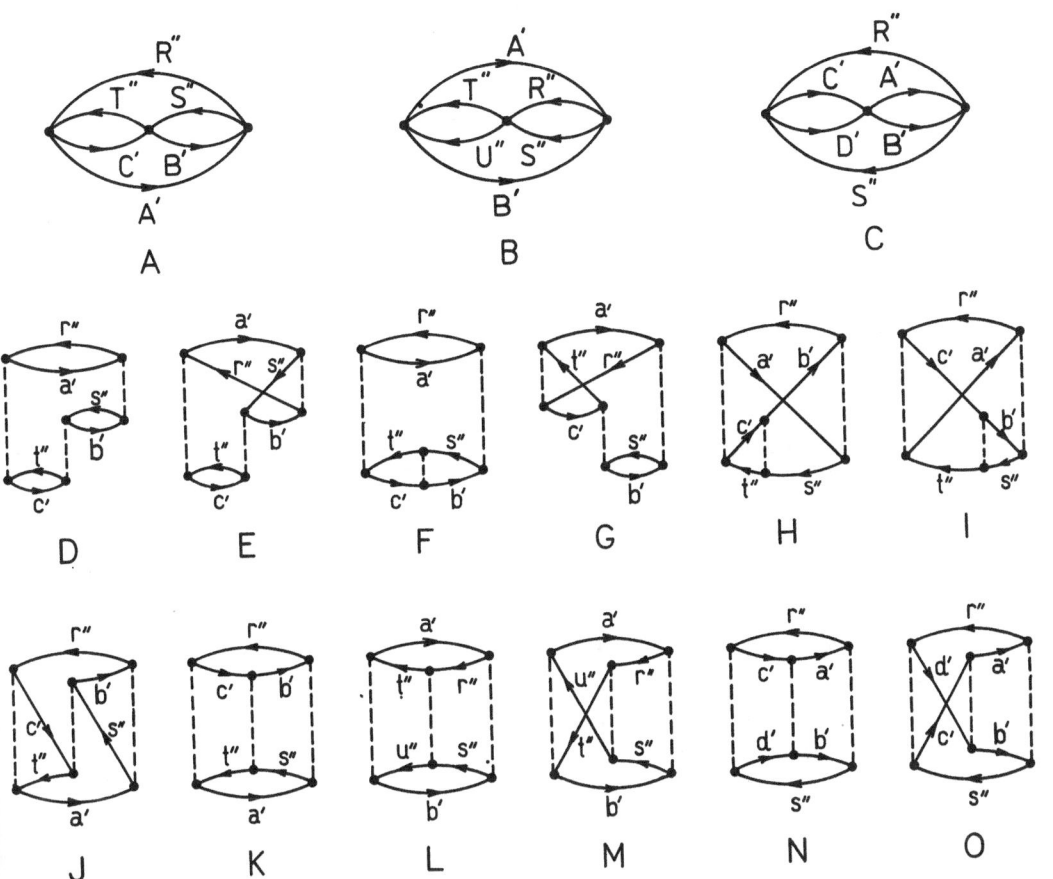

Figure 4.4
Third-order Hugenholtz (A, B, C) and Goldstone (D-O) diagrams.

dices given at oriented full lines. Singly primed indices refer to orbitals occupied in the gound state (hole states) and doubly primed
indices to orbitals unoccupied in the ground state (particle states).
Consistently with previous sections, capital and lowercase letters are
used as indices for spinorbitals and orbitals, respectively. Full
lines oriented from left to right are associated with singly primed
indices and they are therefore called hole lines. Full lines oriented
from right to left are associated with doubly primed indices and they
are called particle lines. The dashed lines in Goldstone diagrams,
that are called the interaction lines, represent the perturbation operator and they entail two-electron integrals according to the following
rule

$$W \longrightarrow \quad \text{or} \quad \text{or} \quad \quad \text{leads to } (ij|kl) \qquad (4.85)$$

so that the indices i and k refer to lines leaving the vertex and the indices j and l refer to lines entering the vertex. The integral is defined as

$$(ij|kl) = \iint \varphi_i(1)\,\varphi_j(1)\,\frac{1}{r_{12}}\,\varphi_k(2)\,\varphi_l(2)\,d\tau_1 d\tau_2 \qquad (4.86)$$

where φ_i, φ_j, φ_k and φ_l are molecular orbitals. Each pair of neighboring interaction lines generates also the denominator factor, generated by Q_0 in eqns. (4.81)-(4.83), R_{ijkl}, which contains a sum of orbital energies ε_i, ε_j, ε_k and ε_l. For hole lines the orbital energies are taken with the plus sign whereas for particle lines they are taken with the minus sign:

$$\frac{1}{\varepsilon_i + \varepsilon_k - \varepsilon_j - \varepsilon_l} \qquad (4.87)$$

For example, for the left half of the diagram D in Fig. 4.4, the denominator factor is

$$R_{a'c'r''t''} = \frac{1}{\varepsilon_{a'} + \varepsilon_{c'} - \varepsilon_{r''} - \varepsilon_{t''}} \qquad (4.88)$$

and for the whole diagram D the denominator factor becomes the product $R_{a'c'r''t}R_{a'b'r''s''}$. The sign corresponding to the algebraic expression of each Goldstone diagram is given by $(-1)^{\ell+h}$ where ℓ is the number of closed loops and h is the number of hole lines. The sum of all possible Goldstone diagrams for the respective order of the perturbation expansion is then multiplied by the topological factor $1/2^n$ where n is the number of equivalent pairs of lines. An equivalent pair consists of two lines which both start at the same vertex of the Hugenholtz diagram and both end at the same vertex. To obtain the expression in the orbital representation, it is necessary to apply the rule for spin summation, viz. to multiply each Goldstone diagram with the factor 2^ℓ.

Prior to passing to the explicit expressions for $E^{(2)}$ and $E^{(3)}$, it is appropriate to make two notes on diagrams in Figs. 4.3 and 4.4: (i) In most of the other literature, the upward and downward orientations of lines are used. We use the orientations from left to right and from right to left in order to preserve the form of diagrams introduced by Paldus and Čížek[320]; (ii) In the literature use is also made of Hugenholtz diagrams drawn in the form of Goldstone-type diagrams but with the meaning of vertices different from that given in eqn. (4.85). Such diagrams will appear later in this section.

Generally speaking, the diagrammatic representation of the respective order of MB-RSPT as given by eqns. (4.81)-(4.83) may be obtained by connecting the hole and particle lines between the vertices with the preserved orientation of lines. It is fair to state, however, that finding all topologically distinct diagrams is by no means easy, especially in higher orders of MB-RSPT. For additional information we refer the reader to the cited literature[238,320]. Once the diagrams are available, their expression is straightforward. For the second-order correlation energy we obtain

$$E^{(2)} = \sum_{\substack{a'\,b'\\r''\,s''}} R_{a'\,b'\,r''s''}(a'\,r''|\,b'\,s'')\left[2(a'\,r''|b'\,s'') - (a'\,s''|b'\,r'')\right] \quad (4.89)$$

The third-order contribution is given by

$$E^{(3)} = E_A^{(3)} + E_B^{(3)} + E_C^{(3)} \quad (4.90)$$

where

$$E_A^{(3)} = \sum_{\substack{a'\,b'\,c' \\ r''\,s''\,t''}} R_{a'\,b'\,r''\,s''} R_{a'\,c'\,r''t''} \left\{ \left[2(a'\,r''\lceil b'\,s'') - (a'\,s''|\,b'\,r'') \right] \right.$$

$$\times \left[2(b'\,s''|\,c'\,t'')\cdot - (b'\,c'|\,s''\,t'') \right] \left[2(a'\,r''|\,c'\,t'') - (a'\,t''|\,c'\,r'') \right]$$

$$\left. - 3(a'\,s''|\,b'\,r'')(b'\,c'|\,s''\,t'')(a'\,t''|\,c'\,r'') \right\} \tag{4.91}$$

represents the contributions from the diagram A in Fig. 4.4 (i.e. from Goldstone diagrams D-K),

$$E_B^{(3)} = \sum_{\substack{a'\,b' \\ r''\,s''\,t''\,u''}} R_{a'\,b'\,r''\,s''} R_{a'\,b'\,t''\,u''}(a'\,r''|\,b'\,s'')(r''\,t''|\,s''\,u'')$$

$$\times \left[2(a'\,t''|\,b'\,u'') - (a'\,u''|\,b'\,t'') \right] \tag{4.92}$$

originates from the diagram B (i.e. Goldstone diagrams L and M), and

$$E_C^{(3)} = \sum_{\substack{a'\,b'\,c'\,d' \\ r''\,s''}} R_{a'\,b'\,r''\,s''} R_{c'\,d'\,r''\,s''}(a'\,r''|\,b'\,s'')(a'\,c'|\,b'\,d')$$

$$\times \left[2(c'\,r''|\,d'\,s'') - (d'\,r''|\,c'\,s'') \right] \tag{4.93}$$

is due to the diagram C (Goldstone diagrams N and O).

The forms of eqns. (4.89)-(4-93) suggest that the evaluation of the second-order contribution is simple and that even the evaluation of the third-order contribution should not be associated with difficulties (see Section 4.K.). Since these simple methods may find wide use, it is topical to accumulate more experience about the correlation energies so obtained. A series of papers oriented to this goal was published for example by Bartlett, Silver and Wilson (see e.g. Refs. 340-342) and by Freemen and Karplus[343], who compared the second- and third-order correlation energies with the estimated "experimental" correlation energies. Testing may also be made by comparing the sec- ond- and third-order MB-RSPT values with the data given by other methods, particularly with the CI-SD data obtained with the same basis set. We present such a comparison[344] for H_2O treated with four different basis sets in Table 4.6 and for HF, Ne and BH in Table 4.7.

T a b l e   4.6

Valence shell correlation energy (E/E_h) in H_2O[a] given by MB-RSPT treatment[344] and the INO-CI calculations[290] including all singly and doubly excited configurations (CI-SD)

Basis set[b]	SCF	CI-SD	$E^{(2)}$	$E^{(2)}+E^{(3)}$	$\dfrac{E^{(2)}+E^{(3)}}{CI\text{-}SD}$ %
[4s2p/2s]	-76.009294	-0.1257	-0.1251	-0.1262	100.4
[4s2p/2s1p]	-76.033838	-0.1393	-0.1388	-0.1408	101.1
[4s2p1d/2s]	-76.036678	-0.1839	-0.1841	-0.1892	102.9
[4s2p1d/2s1p]	-76.048764	-0.1930	-0.1934	-0.1990	103.1

a The geometry assumed: R_{OH} = 1.8089 a_0, ϑ = 104.52⁰.
b For details see Ref. 290.

T a b l e   4.7

Comparison of valence shell (E/E_h) correlation energies in BH, HF and Ne given by MB-RSPT[344], CEPA and CI calculations including all singly and doubly excited configurations (CI-SD)

System	SCF	CI-SD	CEPA	$E^{(2)}$	$E^{(2)}+E^{(3)}$	$\dfrac{E^{(2)}+E^{(3)}}{CI\text{-}SD}$ %
BH[a]	-25.105638	-0.0694[b]	-0.0721	-0.0477	-0.0621	89.5
HF[c,d]	-100.048548	-0.2171	-0.2257	-0.2263	-0.2252	103.7
Ne[c]	-128.524067	-0.2094	-0.2149	-0.2157	-0.2145	102.4

a R_{BH} = 2.336 a_0, [4s2p/2s1p] basis set used (for details see Ref. 345), CI-SD and CEPA results from Ref. 345.
b The full-CI value[345] is -0.0727.
c [5s3p1d/3s1p] basis set used (for details see Ref. 118), PNO-CI and CEPA-PNO results from Ref. 118.
d R_{HF} = 1.733 a_0.

The entries in Tables 4.6 and 4.7 are only valence shell correlation energies but no restriction was imposed upon the number of used unoccupied orbitals given by the particular basis set. From Tables 4.6

and 4.7 it is seen that the correlation energy of ten-electron systems included in $E^{(2)}$ and $E^{(3)}$ is about 100%, compared to CI-SD and CEPA calculations with the same basis set. As expected, the $E^{(2)}$ and $E^{(3)}$ contributions are slightly basis set-dependent, the percentage being larger for the latter than for the former. For the six-electron system BH, $E^{(2)} + E^{(3)}$ gives about 90% of the CI-SD correlation energy and the $E^{(3)}$ contribution in absolute value is considerably larger than it is with the ten-electron systems. We shall not comment on the comparison of $E^{(2)} + E^{(3)}$ with CI-SD from the theoretical point of view (this will be analyzed later on) but rather we notice the fact that the convergence of the perturbation expansion is considerably slower for BH than for the ten-electron species. Hence, if the calculations are restricted to the second and third orders, such a differing convergence may lead to an unbalanced treatment of different systems. We note here two expressions the use of which was suggested for avoiding this effect of truncation of the perturbation expansion at the third order. The first of them is the Padé's [2/1] approximant

$$E[2/1] = E^{(2)}/(1 - E^{(3)}/E^{(2)}) \tag{4.94}$$

Its various forms have been discussed by Wilson et al.[346] (see also Refs. 347,348). The second formula

$$E_V = \frac{E^{(3)} - E^{(2)}}{2S} - \left[\left(\frac{E^{(3)} - E^{(2)}}{2S}\right)^2 + \frac{(E^{(2)})^2}{S}\right]^{1/2} \tag{4.95}$$

is the simplest among the expressions originated from the first-order wave function, $\Phi^{(1)}$ (see e.g. Ref. 349)

$$\Psi = \Phi_o + \gamma \Phi^{(1)} \tag{4.96}$$

with γ determined variationally. For the evaluation of E_V in (4.95) it is necessary besides $E^{(2)}$ and $E^{(3)}$ to compute also

$$
\begin{aligned}
S &= \langle \Phi^{(1)} | \Phi^{(1)} \rangle \\
&= \sum_{\substack{a' b' \\ r'' s''}} R^2_{a' b' r'' s''} (a'\, r'' | b'\, s') \left[2(a'\, r'' | b'\, s'') - (a'\, s'' | b'\, r'')\right]
\end{aligned} \tag{4.97}
$$

Evaluation of S requires, however, no additional computational effort because S differs from $E^{(2)}$ only by the power in the denominator.

The utility of the theory up to the third order has been demonstrated by a series of calculations of molecular geometries, dissociation energies and energy differences between the states of different multiplicity[253], heats of reactions and equilibrium constants of gas-phase reactions[180,350], spectroscopic constants and potential curves [351,352]. Agreement with experiment was substantially better than that achieved by the Hartree-Fock theory.

Higher orders of MB-RSPT

Slow convergence of the perturbation expansion with BH noted above suggests that the higher orders may be important in some cases. For the purposes of the forthcoming discussion on this topic, it is profitable first to make a comparison of the perturbation expression with the CI and cluster expansions and to try to assess what sort of electron correlation is compatible with the particular order of the perturbation expansion. Use may be made of a simple rule which permits expression of the diagrammatic representation of a particular order of the perturbation expansion in terms of the traditional CI: the diagram represents a n-excited configuration if the section between two vertices runs just across n hole lines. Hence, from Figs. 4.3 and 4.4 and Table 4.8 it follows that the second and third orders cover only dou-

T a b l e 4.8

Lowest order (LO) of the perturbation expansion of energy in which various linked and unlinked clusters first appear[323]

Linked clusters	LO	Unlinked clusters	LO
T_1	4	$T_1 T_2$	5
		$\frac{1}{2} T_1^2$	6
T_2	2	$\frac{1}{2} T_2^2$	4
T_3	4	$T_1 T_3$	6

bly excited configurations. Formula (4.89) implies that the second
order accounts only for pair correlation effects. Expressions (4.91)
and (4.93) cannot be split to pair contributions, which means that
the true many-body effects are first met at the third order. Although
the whole effect is due to doubly excited configurations, it is not
saturated at the third order and contributions of doubles appear still
in the fourth and higher orders. (Nevertheless, experience shows that
their second-order contribution is dominant, at least with molecules
in equilibrium geometries). In the fourth order of the perturbation
expansion several new contributions first appear: these are due to singly and triply excited configurations and certain types of quadruple
excitations. In terms of CPMET, these contributions are due to linked
clusters T_1 and T_3 and unlinked clusters $1/2\ T_2 T_2$.

As we already know from the foregoing sections, the dominant part
of the correlation energy is due to doubly excited configurations. It
is therefore natural that attempts were made to develop such a procedure, which would pick up the contributions of doubles from the higher
orders of MB-RSPT. A clear diagrammatic representation of these contributions is possible[354] if use is made of the normalized two-particle matrix elements introduced by Brueckner and Levinson[353]. Application of this approach was reported by Kvasnička and Laurinc[355]. Within the frame of CPMET, the method is equivalent to Čížek's linear approximation[317,354]. The correlation energy corresponding to doubly excited configurations is given by

$$E_D = \qquad\qquad\qquad\qquad\qquad\qquad\qquad\qquad\qquad\qquad (4.98)$$

It holds

$$E_D = \sum_{\substack{a'\,b' \\ r''\,s''}} R_{a'\,b'\,r''\,s''} \left[2(a'\,r''\,|\,b'\,s'') - (a'\,s''\,|\,b'\,r'') \right] (a'\,r''\,|\,g\,|\,b'\,s'') \qquad (4.99)$$

where the matrix element $(a'\,r''\,|g|\,b'\,s'')$ corresponds to a heavy-dot ver-
tex in eqn. (4.98). This vertex is defined diagrammatically[354,355] as
follows

$$(4.100)$$

and the matrix element expressed over spatial orbitals has the following form

$$(a'\, r''|g|b'\, s'') = (a'\, r''|b'\, s'') + \sum_{t''\, u''} R_{a'\, b'\, t''u''}\,(r''\, t''|s''\, u'')(a'\, t''|g|b'\, u'')$$

$$+ \sum_{c'\, d'} R_{c'\, d'\, r''\, s''}\,(a'\, c'|b'\, d')(c'\, r''|g|d'\, s'')$$

$$+ \frac{1}{2}\left[z(a'\, b', r''s'') + z(b'\, a', s''\, r'') \right] \qquad (4.101)$$

where

$$z(a'\, b', r''\, s'') = -2 \sum_{c'\, t''} R_{b'\, c'\, r''\, t''}\,(a'\, c'|s''\, t'')(c'\, r''|g|b'\, t')$$

$$-2 \sum_{c'\, t''} R_{b'\, c'\, s''\, t''}\,(c'\, t''|a'\, r'')(c'\, s'|g|b'\, t'')$$

$$+4 \sum_{c'\, t''} R_{b'\, c'\, s''\, t''}\,(c'\, t''|a'\, r'')(c'\, t''|g|b'\, s'')$$

$$-2 \sum_{c'\, t''} R_{b'\, c'\, s''\, t''}\,(a'\, c'|r''\, t'')(c'\, t''|g|b'\, s'') \qquad (4.102)$$

Equations (4.101) and (4.102) may be solved iteratively with the starting step $(a'\, r''|g|b'\, s'') = (a'\, r''|b'\, s'')$. One may find easily that for the zeroth and first iterations, respectively, the formula (4.99) leads to the second- and third-order expressions of MB-RSPT. In applications to higher orders, the evaluation of $(a'\,r''\,|g|\,b'\,s'')$ in each iteration is evidently a time-consuming step. In the computational scheme suggested by Bartlett, Silver and Shavitt[347,356], the time-consuming step is only involved in each odd order of the perturbation expansion. We preferred, however, to present here the approach based on eqn. (4.98) for its conceptual simplicity. For an example of application we selected the problem of BH (presented in Table 4.7). When the computed $E_D^{(4)} = -0.0057\ E_h$, $E_D^{(5)} = -0.0027\ E_h$ and $E_D^{(6)} = -0.0014$ are added to the third-order energy $E^{(3)}$ (see Table 4.7), the sum of all contributions through the sixth order becomes $-0.0720\ E_h$. Comparing this result with the CI-SD value, we see that the treatment through the third order gives underestimated correlation energy owing to a slow convergence of contributions coming from doubly excited configurations.

There are also other possibilities of summing up certain types of diagrams to infinity. This concerns so called ladder diagrams[331,357] and ring diagrams[358]. The most widely applied approach is the denominator shift technique[340,359] which is based on the Kelly's ladder technique[331,357]. Starting from the second[360] or third order, its implementation is achieved simply by "shifting" the denominator, which means that the denominator appearing in formulas (4.89) and (4.91)-(4.93) contains now besides orbital energies also diagonal elements of the type $(r''\,r''\,|s''\,s'')$, $(a'\,a'\,|\,b'\,b')$ and $(a'\,a'\,|\,r'\,r'')$. It may be shown[361] that this denominator shift leads to the same expression for the correlation energy as if the Epstein-Nesbet[362,363] partitioning of the Hamiltonian is used instead of the Møller-Plesset one. It should be noted that the denominator shift does not mean summing up all double excitations to infinity (as it is, in principle, achievable in the approach discussed in the last paragraph) but only certain types of them (more precisely, the diagonal elements of all ladder and ring diagrams to all orders). Applications of the denominator shift have met with the varying degree of success. Sometimes the results are good and the use of the denominator shift is advocated[348], in other cases they are worse[342,359] than those given by a mere third-order treatment. Application of the denominator shift should definitely be avoided at the second order, because the absence of some third-order terms leads to unbalanced results.

Let us now examine the extension of the explicit treatment to higher orders of the perturbation expansion. This brings about new problems that are not involved in the second and third-order treatments. Through the third order, the classical RSPT and MB-RSPT give the same formulation of the problem. Equivalence order by order is also met in all higher orders, provided that all contributions of the particular order are included. This is, however, hardly achievable for technical reasons, so that much caution is needed in incomplete treatments. We shall concentrate on the fourth order of the perturbation expansion. In this field a remarkable progress has been achieved recently in several laboratories[296,325,326,335,336,364-370].

The fourth-order RSPT energy is given by the expression (4.83), which consists of two terms. The first one, the leading (direct) term involves:

(i) contribution from doubly excited configurations, $E_D^{(4)}$. (Hereafter we shall refer to the second- and third-order energy, $E^{(2)}$ and $E^{(3)}$, as $E_D^{(2)}$ and $E_D^{(3)}$, respectively). This contribution may be evaluated by means of the relationship (4.98).

(ii) contributions from singly and triply excited configurations.

(iii) contributions from quadruply excited configurations. Subsequently we shall consider only the contributions (i) and (iii) which may be assumed to represent the dominant part of the total fourth-order contribution, though there is some evidence[325,369] that the effect of singly and triply excited configurations is not negligible.

The second term in eqn. (4.83) is usually referred to as the renormalization term. It is a product of $E_D^{(2)}$ with the normalization term for the first-order wave function, S, given by eqn. (4.97). Since both $E_D^{(2)}$ and S depend linearly on the number of electrons (see Ref. 371), N, their product $E_D^{(2)}S$ is N^2-dependent and it therefore has to be cancelled.

Again, it is advantageous to introduce the diagrams. The quadruple excitation diagrams of the fourth order are presented in Fig. 4.5. To facilitate the discussion, however, decoding of these diagrams[336] is somewhat different from that used in previous paragraphs: instead of orbitals the summation indices refer to spinorbitals and instead of the integral (4.86) the vertex (4.85) now has the meaning of the antisymmetrized integral

$$(IJ \| KL) = (IJ|KL) - (IL|KJ) \tag{4.103}$$

where $(IJ|KL)$ conforms to the definition (4.86) but is expressed over spinorbitals. Adoption of the definition (4.103) means that all diagrams disappear that are generated by the exchange according to eqn.

116

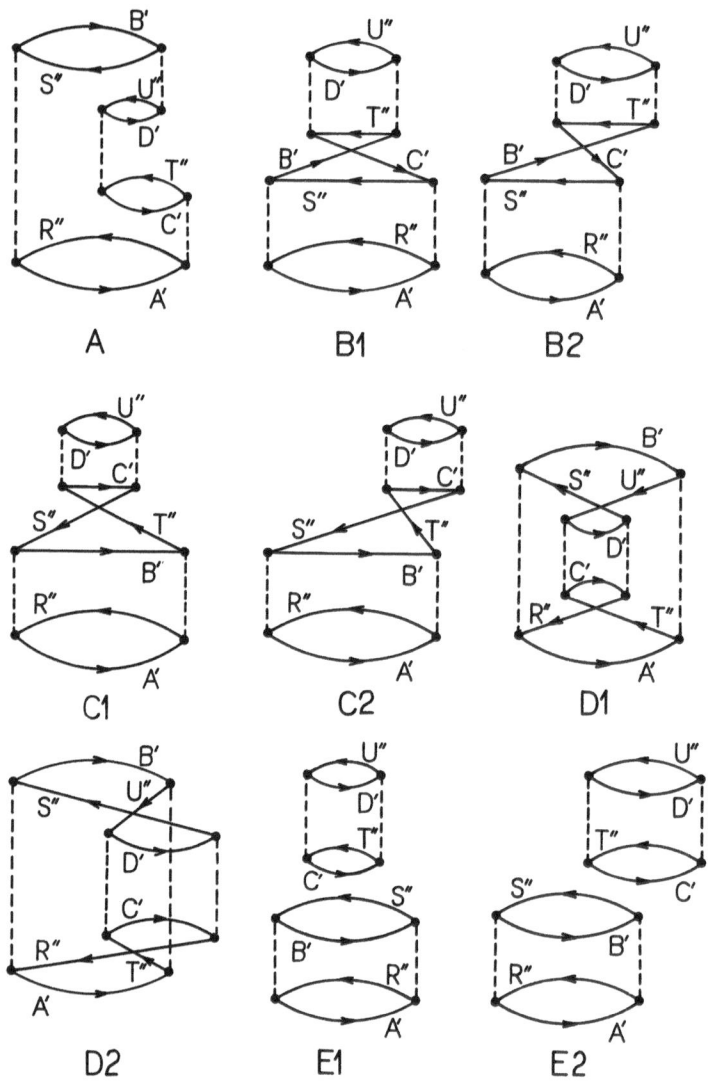

Figure 4.5
Quadruple excitation diagrams which contribute to
the fourth order of MB–RSPT

(4.84). The rules given previously for the denominator, sign and the
topological factor remain unchanged. Using this scheme, the second-order energy, for example, is given by the diagram B in Fig. 4.3 and its
algebraic expression becomes

$$E_D^{(2)} = \frac{1}{4} \sum_{\substack{A' B' \\ R'' S''}} R_{A' B' R'' S''} (A' R'' \| B' S'')^2 \tag{4.104}$$

After the spin summation one finds easily that this expression is equivalent to eqn. (4.89).

As may be noticed in Fig. 4.5, the diagrams are so labelled to imply certain pairing of diagrams. This pairing has its origin in the effective algebraic evaluation of diagrams. For example

$$D1 + D2 = \frac{1}{16} \sum_{\substack{A' B' C' D' \\ R'' S'' T'' U''}} R_{A' B' R'' S''} R_{A' B' T'' U''} R_{C' D' R'' S''}$$

$$\times (A' R'' \| B' S'')(C' T'' \| D' U'')(C' R'' \| D' S'')(A' T'' \| B' U'') \tag{4.105}$$

For the manipulation with denominators use was made of the identity $1/X(X + Y) + 1/Y(X + Y) = 1/XY$. Among the diagrams A - E in Fig. 4.5, the first seven ones are referred to as the connected diagrams, whereas the remaining two are referred to as the disconnected diagrams. As we shall see, understanding of relations between the connected diagrams, disconnected diagrams and the $E_D^{(2)}S$ term is essential for a correct fourth-order treatment of double and quadruple excitations. First we decode the E diagrams which gives us the following important expression

$$E1 + E2 = \frac{1}{16} \sum_{\substack{A' B' C' D' \\ R'' S'' T'' U''}} \sum R_{A' B' R'' S''}^2 R_{C' D' T'' U''} (A' R'' \| B' S'')^2$$

$$\times (C' T'' \| D' U'')^2 = E_D^{(2)}S \tag{4.106}$$

As we already know both $E_D^{(2)}$ and S are related to double excitations. $E_D^{(2)}$ was discussed in detail in this section and S was shown in eqn. (4.97) to be given by the coefficients of perturbation contributions to the first-order wave function

$$\Phi^{(1)} = \frac{1}{4} \sum_{\substack{A' B' \\ R'' S''}} R_{A' B' R'' S''} (A' R'' \| B' S'') \Phi_{A' B'}^{R'' S''}$$ (4.107)

so that

$$S = \frac{1}{4} \sum_{\substack{A' B' \\ R'' S''}} R_{A' B' R'' S''}^2 (A' R'' \| B' S'')^2$$ (4.108)

Equation (4.106) leads directly to the conclusion that the energy in the fourth order of MB-RSPT is given solely by the connected diagrams A-D. This finding is, of course, just what follows from the formulation of the linked cluster theorem[339,372] in MB-RSPT. It guarantees automatically that the size-inconsistent term $E_D^{(2)}$S is cancelled by means of contributions from disconnected quadruple excitation diagrams E1 and E2. (Note the minus sign standing at the renormalization term $E_D^{(2)}$S in eqn. (4.83).) The contributions from diagrams A-D are due to disconnected wave function clusters of double excitations, $1/2 \ T_2 T_2$ (the terminology of MB-RSPT has not yet become unequivocal; the terms "connected" and "disconnected" used by Bartlett and Purvis[326a] have the same meaning as terms "linked" and "unlinked" used in Sections 4.F. and 4.G.).

Having analyzed the correlation energy contributions from diagrams A-E, the next point of examination is the case when coincidence of indices occurs for any pair of particle or hole lines. Such contributions are denoted in the literature with the acronym EPV (exclusion principle violating). In the language of creation and annihilation operators it means that a particle or hole is created twice before it is annihilated which is why the "exclusion principle violating" designation arose. Assume, for example, that indices B' and C' in the diagram A are the same. Fig. 4.6 shows us that a modified linkage of the hole lines B' and C' gives us the diagram C1. Apart from the sign, the two diagrams yield the same expression, which brings about the cancellation of the EPV contribution. Next we may assume, for example, the e-quality of S" and T" in the diagram C1. Here a modified linkage of lines leads to the diagram E1. Hence, the cancellation of this EPV contribution is brought about by the interplay between connected and disconnected diagrams. A complete analysis along these lines[280,373] gives

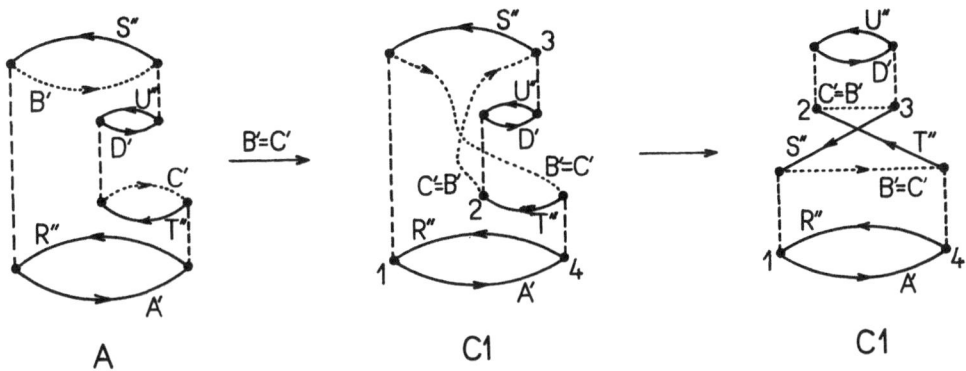

Figure 4.6
An example of the cancellation of EPV contributions

us an important conclusion viz. that the cancellation of EPV contributions within the context of diagrams A-D is incomplete. A part of
their EPV contributions is cancelled by the EPV contributions from the
diagrams E1 and E2 giving

$$(A + B + C + D)_{EPV} = -(E)_{EPV} \qquad (4.109)$$

so that

$$\sum (A \text{ through } E)_{EPV} = 0 \qquad (4.110)$$

The knowledge we acquired from the derivation of eqns. (4.106) and
(4.110) is embodied in Table 4.9. It is seen that the only nonvan-
ishing terms in this approach are the NEPV contributions from connected diagrams and "EPV" contributions from $-E_D^{(2)}$s. The non-EPV part
of the $-E_D^{(2)}$s has to be cancelled out because it brings about size-in-
consistency. On the other hand, the "EPV" part of $-E_D^{(2)}$s is linearly
dependent on the number of electrons and it is preserved. Of course,
this term by no means violates the exclusion principle. It is only due
to double excitations and it has a correct physical meaning. Instead
of EPV it should be rather referred to as the "conjoint" contribution
[280,326,365]. On the contrary, the disjoint term ("NEPV") is unphysical
and it must be removed in any theory in which the effect of double ex-

citations is rigorously included.

T a b l e 4.9
Contributions of double and quadruple excitations in the fourth order of MB-RSPT. NEPV means non-EPV contributions. The boxed entries are contributions that cancel out

Origin	Type of the contribution	
Connected diagrams (A-D in Fig. 4.5)	EPV	NEPV
Disconnected diagrams (E in Fig. 4.5)	EPV	NEPV
$-E_D^{(2)}S$ term	"EPV"	"NEPV"

We may now apply relationships (4.106) and (4.109) to the two uncancelled terms in Table 4.9. This gives us the following result

$$(\text{A through D})_{NEPV} - (E_D^{(2)}S)_{"EPV"} = (\text{A through D})_{NEPV} - (E)_{EPV}$$

$$= (\text{A through D})_{NEPV+EPV} \qquad (4.111)$$

which is again in accordance with the linked cluster theorem.

Separation to EPV and NEPV contributions is of great importance for finding approximate relations between different methods. Listed below are the quantities that permit us a deeper insight into the problem. All these quantities may be taken as contributions to the correlation energy given by MB-RSPT through fourth order.

(1) Sum of contributions due to doubly excited configurations,

$$E_D^{(2)-(4)} = E_D^{(2)} + E_D^{(3)} + E_D^{(4)} \qquad (4.112)$$

which represents the contribution of double excitations from direct terms. It may be compared to the energy given by Čížek's linear approximation[317,354,374] in CPMET.

(2) The relationship,

$$E_{CI-D}^{(2)-(4)} = E_D^{(2)-(4)} - E_D^{(2)}S \qquad (4.113)$$

may be taken as the approximation to CI covering doubly excited configurations. The term $-E_D^{(2)}S$ is inherently positive. In the CI-D approach it brings about size inconsistency, but it ensures the upper
bound to energy[336], a typical property of variational methods. Of
course, CI-D includes the effect of double excitations up to the "infinite" order. This coverage is, in principle, achievable in the perturbation treatment by means of eqn. (4.98), but it would be necessary
to augment eqn. (4.113) with additional renormalization terms for
higher orders. For example, the supplementary terms for the fifth or-
der are $-E_D^{(3)}S$ and $-2E_D^{(2)} \langle \Phi^{(2)} | \Phi^{(1)} \rangle$, where $\Phi^{(2)}$ is the second-
-order wave function. A thorough analysis of the effect of such sup-
plementary terms was reported by Siegbahn[296]. The problem noted in
this paragraph is general and it is inherent to any method in which
higher-order terms $E_D^{(n)}$ are involved.

(3) Within the frame of MB-RSPT through the fourth order, a rigorous expression for the contribution from doubly excited configurations
becomes

$$E_{DR}^{(2)-(4)} = E_D^{(2)-(4)} - (E_D^{(2)}S)_{"EPV"} \qquad (4.114)$$

The subscript R is used here for pointing out that the second term in
eqn. (4.114) comes from the fourth-order renormalization term. Equations (4.113) and (4.114) show clearly differing meanings of the concept "effect of double excitations" in CI-D and MB-RSPT.

(4) The contribution coming from quadruply excited configurations
is as follows

$$E_Q^{(4)} = (A \text{ through } D)_{NEPV} \qquad (4.115)$$

(5) The fourth-order contribution including both doubly and quadruply excited configurations is given by

$$E_{DQR}^{(4)} = E_D^{(4)} + E_Q^{(4)} - (E_D^{(2)}S)_{"EPV"}$$

$$= E_D^{(4)} + (A \text{ through } D)_{EPV+NEPV} = E_D^{(4)} + E_{QR}^{(4)} \qquad (4.116)$$

(6) A complete expression for the correlation energy through

fourth order in the considered MB-RSPT approach including doubly and quadruply excited configurations, is given by

$$E_{DQR}^{(2)-(4)} = E_{D}^{(2)-(4)} + E_{QR}^{(4)} \tag{4.117}$$

(7) For the normalized form of the wave function given by the first order perturbation theory

$$\Psi = (1 + S)^{-1/2}(\Phi_0 + \Phi^{(1)}) \tag{4.118}$$

we may assume that it approximates the CI wave function truncated to double excitations

$$\Psi_{CI-D} = c_0\Phi_0 + \sum_{A>B} \sum_{R>S} c_{AB}^{RS} \Phi_{AB}^{RS} \tag{4.119}$$

From eqn. (4.119) we obtain

$$\sum_{A>B} \sum_{R>S} (c_{AB}^{RS})^2 = 1 - c_0^2 \tag{4.120}$$

and the comparison of coefficients standing at doubly excited configurations in eqns. (4.118) and (4.119) (see also eqns. (4.107) and (4.108)) gives us

$$1 - c_0^2 = \frac{S}{1 + S} \tag{4.121}$$

Assuming next $E_D^{(2)} \approx E_{CI-D}$ we may write to first order in S

$$(1 - c_0^2)E_{CI-D} \approx E_D^{(2)}S \tag{4.122}$$

Thus the $E_D^{(2)}S$ term may be taken as an approximation to Davidson's formula (4.19) which corrects a posteriori the size inconsistency of the CI-SD wave function[296,336]. Comparison with $E_{DR}^{(2)-(4)}$ suggests that it neglects the term $-(E_D^{(2)}S)_{"EPV"}$ which should be involved. A reasoning usually given to Davidson's formula is that it accommodates approximately quadruply excited configurations. According to Table 4.9 this should be understood as implicit inclusion of those contributions from quadruply excited diagrams, that furnish the cancellation of "NEPV" contributions and also (erroneously) of "EPV" contributions from the $-E_D^{(2)}S$ term. Contributions corresponding to $E_Q^{(4)}$ are by no means involved, however.

(8) EPV contributions also permit us to find relationships between MB-RSPT and CEPA. It may be shown that differences in several existing variants of CEPA may be assigned just to the differing extent of inclusion of EPV contributions. A detailed discussion on this problem may be found, for example, in papers by Kutzelnigg[319], Ahlrichs[298] and Hurley[280]. A diagrammatic treatment was presented by van der Velde[16] and Robb[335]. We restrict here ourselves to a brief summary:

(i) in all variants of CEPA the $-(E_D^{(2)}S)_{"NEPV"}$ term and NEPV contributions from diagrams E are correctly cancelled. The $-(E_D^{(2)}S)_{"EPV"}$ term is taken in an approximate way. In this respect it is possible to speak of the implicit inclusion of quadruple excitations in CEPA.

(ii) double excitations are summed up to infinite order.

(iii) NEPV contributions from the diagrams A through D are not recognized so that the energy $E_Q^{(4)}$ is not involved in CEPA explicitly.

(iv) CEPA also permits inclusion of singly excited configurations.

To illustrate the quantities discussed in previous paragraphs we present some numerical results[366,367]. We note first the entries of Table 4.10 for the water molecule, because in this case we have CI data available that are suitable for the purposes of our discussion. We see that among the fourth-order contributions, $E_D^{(4)}$, $E_Q^{(4)}$, and $-(E_D^{(2)}S)_{"EPV"}$, the first two are negative and the third one is positive but that they are roughly equivalent in absolute value. Ratios between these three quantities may be viewed in two different ways:
(a) Double excitation terms $E_D^{(4)}$ and $-(E_D^{(2)}S)_{"EPV"}$ approximately cancel out so that the resulting total double-excitation contributions, $E_{DR}^{(2)-(4)}$, is given almost solely by $E_D^{(2)} + E_D^{(3)}$. This means that $E_Q^{(4)}$ is a dominant fourth-order contribution. Alternatively, (b) we may consider an approximate cancellation of terms $E_Q^{(4)}$ and $-(E_D^{(2)}S)_{"EPV"}$. According to eqn. (4.111) the sum of these two terms is obtained by summing up both EPV and NEPV contributions from the diagrams A-D. This

T a b l e 4.10

Correlation energy contributions[367] given by MB–RSPT through
fourth order for H_2O with the contracted Gaussian DZ basis set[a]

Energy contribution[b]		E/E_h
E_{SCF}		-76.00929
Double excitation terms:	$E_D^{(2)}$	-0.12510
	$E_D^{(3)}$	-0.00111
	$E_D^{(4)}$	-0.00293
	$E_D^{(2)}S$	-0.00504
	$(E_D^{(2)}S)_{"EPV"}$	-0.00257
Quadruple excitation term :	$E_{QR}^{(4)}$	-0.00081
Various quantities:	$E_D^{(2)-(4)}$	-0.12914
	$E_{CI-D}^{(2)-(4)}$	-0.12410
	$E_{DR}^{(2)-(4)}$	-0.12657
	$E_Q^{(4)}$	-0.00338
	$E_{DQR}^{(4)}$	-0.00374
	$E_{DQR}^{(2)-(4)}$	-0.12995
	$E_Q^{(4)} + (E_D^{(2)}S)_{"NEPV"}^{c}$	-0.00586
CI data[7]:	E_{CI-SD}	-0.12615
	$E_{CI-SDTQ} - E_{CI-SDT}$	-0.00577

[a] For details on the basis set and geometry used see Refs. 7
    and 367.
[b] For definitions see text.
[c] The MB–RSPT analog of the difference $E_{CI-SDTQ} - E_{CI-SDT}$.

sum, which is referred to as $E_{QR}^{(4)}$, represents therefore the rest contribution after an incomplete cancellation of $E_Q^{(4)}$ and $-(E_D^{(2)}S)_{"EPV"}$.
Ordinarily, $E_{QR}^{(4)}$ is small[365] and positive (the negative value in Table
4.10 is an artefact of the small basis set).

Cancellation in the sense of item (b) has an interesting consequence. If $E_{QR}^{(4)}$ turned out to be generally small, then double excitation contributions would be dominant. If it were so, we might also regard those procedures as rigorous that disregard in fourth order both

quadruply excited configurations and "EPV" terms from $E_D^{(2)}S$. Actually, in the perturbation treatment it would be sufficient to assume the term $E_D^{(2)-(4)}$ which is obtainable in a relatively simple way. In the CI approach, the above noted cancellation would mean numerical (a posteriori) justification of Davidson's correction. The last entry listed in the family of "various quantities" in Table 4.10, mimics the effect of augmenting the CI wave function with the quadruply excited configurations. Agreement between the fourth-order MB-RSPT and CI is remarkable on this point. It should be realized, however, that in MB-RSPT a part of disconnected T_2T_2 clusters (those from higher orders) is missing. Connected clusters are completely missing and their contributions appear only in higher orders. In contrast, the CI-SDTQ energy contains additional renormalization terms which should be cancelled out by higher configurations.

Table 4.11 presents the results for Ne, HF and H_2O. In contrast to Table 4.10, where use was made of the DZ basis set, $E_{QR}^{(4)}$ given by the DZ+P basis set is positive in all three cases listed in Table 4.11. It is noteworthy that in the series Ne-HF-H_2O the cancellation of $E_Q^{(4)}$ and $-(E_D^{(2)}S)$"EPV" terms becomes more incomplete. The same trend may be inferred from the data reported by Krishnan and Pople[365]. In absolute value, the $E_Q^{(4)}$ term is small with all three systems assumed (1.7% of $E_D^{(2)-(4)}$ with H_2O, where its percentage is the highest). As the size of the molecule is increased, however, the magnitude of $E_Q^{(4)}$ may rise rapidly[276]. With the N_2 molecule, for example, it becomes[367] -0.01009 E_h, which represents 3.2% of $E_D^{(2)-(4)}$, though it reduces to +0.00568 E_h after the cancellation with the $-(E_D^{(2)}S)$"EPV" term.

Obviously, much more numerical data are needed to assess in general the relative importance of quantities discussed above. Additional information may be found in pioneering papers by Bartlett, Purvis and Shavitt[294,326,364], Pople and coworkers[325,365], Wilson, Silver and Saunders[368,369], and Siegbahn[296]. However, for a larger set of molecules only the results by Bartlett and Purvis[326] permit detailed analysis as it was presented above, although only with approximate $(E_D^{(2)}S)$"EPV" terms. From the data by Krishnan and Pople[365] it is possible to infer $E_{QR}^{(4)}$, but since the EPV contributions cannot be separated, it is not possible to arrive at $E_Q^{(4)}$ by means of our relationship (4.115). Anyway, the reported fourth-order MB-RSPT calculations have contributed much to the understanding of the correlation problem and they showed clearly that the following treatments of other molecules are highly topical.

T a b l e 4.11

Correlation energy contributions[367] (E/E_h) given by
MB-RSPT through fourth order for 10-electron systems
with the DZ+P basis set

Energy contribution[a]	Ne	HF	H_2O
E_{SCF}	-128.52395	-100.04787	-76.04647
$E_D^{(2)}$	-0.17755	-0.19699	-0.19884
$E_D^{(3)}$	-0.00181	-0.00204	-0.00603
$E_D^{(4)}$	-0.00132	-0.00256	-0.00342
$E_D^{(5)}$	-0.00023	-0.00054	-0.00090
$E_D^{(6)}$	-0.00007	-0.00021	-0.00035
$E_D^{(2)}S$	-0.00498	-0.00840	-0.01046
$(E_D^{(2)}S)_{"EPV"}$	-0.00256	-0.00429	-0.00530
$E_{QR}^{(4)}$	+0.00043	+0.00099	+0.00170
$E_D^{(2)-(4)}$	-0.18068	-0.20160	-0.20829
$E_{CI-D}^{(2)-(4)}$	-0.17570	-0.19321	-0.19783
$E_{DR}^{(2)-(4)}$	-0.17812	-0.19732	-0.20299
$E_Q^{(4)}$	-0.00213	-0.00330	-0.00361
$E_{DQR}^{(4)}$	-0.00089	-0.00158	-0.00172
$E_{DQR}^{(2)-(4)}$	-0.18025	-0.20061	-0.20660

[a] For definitions see text.

4.K. Numerical Treatment of Perturbation Expressions

This section was intended to have a similar role in this chapter
to that in Chapter 3. of the Sections 3.C. and 3.D. devoted to problems of time saving in SCF calculations. With the calculations of correlation energy, however, the situation is more complex because of a
great number of different methods among which each has its own specific theoretical problems and, accordingly, it requires special tricks
for accelerating the actual calculations. Since the most widely used
methods, CI-SD and CEPA, have already been discussed in some detail in

previous sections, the emphasis is laid here on MB-RSPT calculations, for which we have acquired some personal experience.

A typical (though not always unavoidable) first step in any procedure for the evaluation of correlation energy is the transformation of integrals over AO's (i.e. STO's, CGTF's and other possible basis set functions) with the indices μ, ν, λ, and σ to integrals over MO's with the indices i,j,k and ℓ:

$$(ij|k\ell) = \sum_{\mu,\nu,\lambda,\sigma} c_{i\mu} c_{j\nu} c_{k\lambda} c_{\ell\sigma} (\mu\nu|\lambda\sigma)$$ (4.123)

If the number of basis set functions is n, the number of $(ij|k\ell)$ integrals becomes $\approx n^4$. Since the evaluation of each of the latter requires the manipulation with $\sim n^4$ integrals $(\mu\nu|\lambda\sigma)$, the direct transformation according to eqn. (4.123) is extremely ineffective. Sutcliffe[375] was probably the first who suggested the $\sim n^5$ dependent algorithm. Essentially, the algorithm contains four steps in which partly transformed integrals are stepwise constructed according to the following scheme:

$$(i\nu|\lambda\sigma) = \sum_{\mu} c_{i\mu}(\mu\nu|\lambda\sigma)$$ (4.124)

$$(ij|\lambda\sigma) = \sum_{\nu} c_{j\nu}(i\nu|\lambda\sigma)$$ (4.125)

$$(ij|k\sigma) = \sum_{\lambda} c_{k\lambda}(ij|\lambda\sigma)$$ (4.126)

$$(ij|k\ell) = \sum_{\sigma} c_{\ell\sigma}(ij|k\sigma)$$ (4.127)

Making use of this algorithm requires of course to store the partly transformed integrals. It is profitable to store them in blocks according to the index i. Development[275,376-379] along these lines resulted in highly effective algorithms.

For second- and third-order calculations only integrals of certain types are needed. Denoting the occupied orbitals by o and virtual orbitals by v, the integrals met in the second-order calculations are only of the (ov|ov) type. For third-order calculations we also need integrals (oo|vv), (oo|oo) and (vv|vv). For quadruply excited configurations in the fourth-order treatments of ground states, only (ov|ov) integrals are needed. All types of integrals, i.e. also of (oo|ov) and (ov|vv) types, are required, if singly and triply excited configurations are included in the fourth-order treatment. Separated blocks may be obtained directly by a suitable sequence of transformations and by storing the partly transformed integrals on intermediate tapes or discs. The second-order energy (4.89) is obtainable readily, if all (ov|ov) integrals are retained in the core, the core requirement being $N_v \times N_o \times (N_v \times N_o + 1)/2$ (double precision) words, where N_v is the number of virtual MO´s and N_o is the number of occupied MO´s assumed in the particular run. It is more advantageous to make use of segments of (ov|ov) integrals, as they were transformed with respect to the first index. With the indices of eqn. (4.89) the number of segments is determined by the index $a´$. Each segment contains integrals $A_{a´} (r´´,b´, s´´)$ and the second-order energy is calculated as

$$E^{(2)} = \sum_{\substack{a´ \ b´ \\ r´´ s´´}} R_{a´ b´ r´´ s´´} A_{a´} (r´´,b´,s´´) \left[2A_{a´} (r´´,b´,s´´) - A_{a´} (s´´,b´,r´´) \right]$$

(4.128)

From the point of view of reducing the memory requirements, one might expect it preferable to interchange indices so that integrals $A_{r´´} (a´,s´´,b´)$ would be stored. From the point of view of economy, however, it is profitable, if in the first step of integral transformation, which is the most time consuming step, the index refers to occupied MO´s (provided $N_o < N_v$). Anyhow with respect to the integral transformation, the time required for the evaluation of $E^{(2)}$ is small.

Evaluation of the third-order energy is somewhat more complex. Consider, for example, the contribution $E_B^{(3)}$ given by eqn. (4.92). It contains integrals of the (vv|vv) type, which by no means can be kept in the memory. A procedure which we use in our program[380] is the following:

(i) Read in a segment from a file containing (vv|vv) integrals.
(ii) Unpack the indices, $r´´,t´´,s´´,u´´$ and find the value, V, with any member of the segment.
(iii) To any integral $V = (r´´t´´|s´´u´´)$ find the corresponding (ov|ov) integrals from the segments $A_{a´}$, evaluate and add the respec-

tive contribution according to the formula

$$E_B^{(3)} = \sum_{V \in (vv|vv)} \sum_{a'} \sum_{b'} R_{a'\,b'\,r''\,s''} R_{a'\,b'\,t''\,u''} V$$

$$\times A_{a'}(r'',b',s'') \left[2A_{a'}(t'',b',u'') - A_{a'}(u'',b',t'') \right] \quad (4.129)$$

(iv) Return to (i) unless the treated block of (vv|vv) integrals was
the last one.

The contributions $E_A^{(3)}$ and $E_C^{(3)}$ are evaluated in a similar way. Additional simplification may be achieved by introducing the denominators
directly into the matrix elements

$$AR_{a'}(r'',b',s'') = (a'\,r''|\,b'\,s'')R_{a'\,b'\,r''\,s''} \quad (4.130)$$

and similarly for $A_{a'}(t'',b',u'')$ and $A_{a'}(u'',b',t'')$. (Note that terms
(4.130) have the meaning of coefficients in the first-order wave function (4.107)). The number of operations in the third-order calculations increases as $N_o^2 N_v^4$, the factor being only moderately higher than
that with SCF calculations. A similar procedure for the calculation of
correlation energy by means of MB-RSPT through third order was reported by Silver and Wilson[381-383]. A remarkable feature of the algorithm by Pople and co-workers[253] is that it avoids the most time-consuming step in the integral transformation - the transformation to
(vv|vv) integrals, and evaluates the $E_B^{(3)}$ terms directly from inte-
grals $(\mu\nu|\lambda\sigma)$ over basis set functions.

Consider next the fourth-order contributions. Since the computational aspects of the evaluation of terms that are due to doubly excited configurations were discussed in some detail in the previous
section, we restrict ourselves here to quadruple excitation diagrams.
From the inspection of eqn. (4.105) it might appear that the eight-
-fold summation makes the evaluation of diagrams D prohibitive. The
same problem is also met with the diagrams A-C. Fortunately, this bottleneck may be eliminated by introducing intermediate arrays. In our
treatments we adopted the formulation by Robb[335]. Essentially the same
technique is also used in other reported algorithms[365,368,384]. The
idea is relatively simple. Note that the indices R'' and S'' in eqn.
(4.105), for example, appear only in the first and third integral. For

this reason it is profitable to define

$$G(A',B',C',D') = \sum_{R''S''} R_{A'B'R''S''} R_{C'D'R''S''} (A'R''\|B'S'')(C'R''\|D'S'') \quad (4.131)$$

where the number of multiplications is proportional to N_v^2 (for each of approximately N_o^4 elements of the G array). Similarly, we may define

$$F(A',B',C',D') = \sum_{T''U''} R_{A'B'T''U''} (C'T''\|D'U'')(A'T''\|B'U'') \quad (4.132)$$

Note that both G and F may be calculated simultaneously within the final summation which gives us the expression for $D_1 + D_2$ as

$$D_1 + D_2 = \frac{1}{16} \sum_{A'B'C'D'} G(A',B',C',D')F(A',B',C',D') \quad (4.133)$$

The most time consuming step (approximately with $N_o^4 N_v^2$ operations) is the formation of G and F. The use of intermediate sums as outlined above permits the number of multiplications with any fourth-order diagram not to exceed the level of third-order treatments. Of course, after spin summations[368a,384] the calculations may be performed in integrals over spatial orbitals, not in integrals over spinorbitals.

The usual way of evaluating the individual EPV and NEPV fourth--order contributions is to compute directly the energy $E_{QR}^{(4)}$, i.e. the contribution from diagrams (A through D)$_{EPV+NEPV}$. The term $-(E_D^{(2)}S)_{"EPV"}$ is either approximated by means of a simple expression with pair correlation energies[280,326,364] or it is calculated rigorously. We have shown recently[367] that neither the rigorous calculation of $-(E_D^{(2)}S)_{"EPV"}$ is very time consuming. The $E_Q^{(4)}$ term is readily determined as the difference $E_{QR}^{(4)} - [-(E_D^{(2)}S)_{"EPV"}]$.

Finally, it is fair to note that the individual contributions $E_Q^{(4)}$ and $-(E_D^{(2)}S)_{"EPV"}$ are not invariant to unitary transformation among degenerate orbitals. Only their sum, $E_{QR}^{(4)}$, is invariant. This does not matter, however, if one uses these contributions for a critical examination of the various approaches in which some of the terms $E_D^{(2)}$, $E_Q^{(4)}$ and $-(E_D^{(2)}S)_{"EPV"}$ are ignored or approximated.

4.L. Basis Set Dependence

At the beginning of this chapter we noted that the capability of
the present computational methods is to give 80% (or more) of correlation energy. This high percentage is, however, attainable only with
very large basis sets. For the water molecule, Meyer[292] obtained with
the PNO-CI and CEPA calculations, respectively, the correlation energies of -0.3056 and -0.3201 E_h, i.e. 86 and 90% of the "experimental"
correlation energy. He used the GTF basis set (11s7p4d1f/5s1p) augmented with a 1s bond function in each bonding region. The largest
among the STO basis sets considered in Table 2.15 can give only 75% of
the correlation energy, though it gives lower SCF energy compared to
Meyer's calculation. Ahlrichs and coworkers[120] tested the effect of polarization functions with the nitrogen molecule. They examined the
gain in SCF and correlation energies upon augmenting the basis set suc-
cessively with the first d-set, second d-set and f-set. They concluded
that the polarization functions which can safely be neglected in SCF
calculations may be very important for the correlation energy. So it
should be kept in mind that not only the number of configurations,
cluster types, or the order of the perturbation expansion, but also
the basis set size is a limiting factor in accounting for the correlation energy.

4.M. Size Consistency

In Section 4.D. we noted on the PNO-CI calculations[250] for the
dimerization of BH_3. It was shown there (eqns. (4.17) and (4.18)) that
different energies of dimerization are obtained if they are computed
as $E(B_2H_6) - 2E(BH_3)$ and $E(B_2H_6) - E(BH_3 \cdots BH_3)$ respectively. A more
realistic ΔE is obtained in the latter case in which one substracts
the energy of a supersystem $BH_3 \cdots BH_3$ with a very large intermolecular
distance from the energy of the dimer. This is an excellent example of
what is referred to as the incorrect dependence on the number of particles.

Size inconsistency of the CI-SD approach was analyzed in Section
4.J., where it was shown that the CI-SD energy contains the $E_D^{(2)}S$ term,
which has the N^2 dependence instead of the correct N dependence. For
those who prefer the traditional CI-formalism to the perturbation
treatment, we present here the demonstration reported by van der Velde[16] and Ahlrichs et al.[293]. Consider a model system of n identical noninteracting two-electron molecules (say H_2 molecules) which is treated

by the ordinary configuration interaction including only doubly excited configurations. First we rewrite eqns. (4.54) and (4.55) in the CI formalism

$$c_0 \langle \Phi_0 | H | \Phi_0 \rangle + \sum_{A<B} \sum_{R<S} c_{AB}^{RS} \langle \Phi_0 | H | \Phi_{AB}^{RS} \rangle = c_0 E \qquad (4.134)$$

$$c_0 \langle \Phi_{AB}^{RS} | H | \Phi_0 \rangle + \sum_{C<D} \sum_{T<U} c_{CD}^{TU} \langle \Phi_{CD}^{TU} | H | \Phi_{AB}^{RS} \rangle = c_{AB}^{RS} E \qquad (4.135)$$

Next we assume that a minimum basis set is used so that only one virtual orbital is available for a single molecule. Then the CI wave function contains n doubly excited configurations, each of them having the same energy (E_1), expansion coefficient (c_1) and the matrix element with the ground state (H_{01}). Assuming this and $\langle \Phi_0 | H | \Phi_0 \rangle = 0$, eqns. (4.134) and (4.135) reduce to

$$E_{corr} = n \frac{c_1}{c_0} H_{01} \qquad (4.136)$$

and

$$c_0 H_{01} + c_1 E_1 = c_1 E_{corr} \qquad (4.137)$$

which gives us

$$\lim_{n \to \infty} E_{corr} = - \sqrt{n} |H_{01}| \qquad (4.138)$$

The obtained $\sqrt{n}$ dependence may appear to be in conflict with the finding that the energy given by the CI-SD wave function contains the $E_D^{(2)}$S term, which is N^2 dependent (N being the number of electrons). Kutzelnigg et al.[371b] showed, however, that the two seemingly different statements are consistent. Equation (4.136) may be expressed[371b] as

$$E_{corr} = \frac{E_1}{2} - \frac{1}{2} \sqrt{E_1^2 + 4nH_{o1}^2} \qquad (4.139)$$

and provided that

$$n < \frac{E_1^2}{4H_{o1}^2} \qquad (4.140)$$

it may be expanded as a Taylor series in n

$$E_{corr} = - \frac{nH_{o1}^2}{E_1} + \frac{n^2 H_{o1}^4}{E_1^3} - 2 \frac{n^3 H_{o1}^6}{E_1^5} - \ldots \qquad (4.141)$$

For a sufficiently small value of nH_{o1}^2/E_1, the deviation from the correct n-dependence occurs in the term proportional to n^2, in agreement with that what we learned in Section 4.J.

From eqn. (4.138) a very important consequence follows viz. that a CI treatment limited to singly and doubly excited configurations cannot be applied rigorously to comparisons of molecules of different size. The CI approach becomes size consistent only if quadruple excitations are involved or if some correction such as Davidson's expression is applied. As regards the other methods noted in this chapter, IEPA, MET, CPMET, CEPA, and various MB-RSPT approaches, they are all size-consistent.

5. Applications

In applications of ab initio calculations to chemical problems, three levels may be distinguished:

(i) Studies belonging to this category furnish the assessment of the reliability of the methods for a particular problem

(ii) The calculations aim at confirming and supplementing the experimental results

(iii) The calculations yield reliable predictions of data that are not obtainable from experiment.

We attempted to survey the knowledge acquired in studies of the category (i) by presenting some typical examples and general trends. These are hoped to be useful for selecting the computational approach that would be compatible with the requirements involved in applications of the types (ii) and (iii).

Before discussing the applications to individual chemical problems, it should be noted that the correspondence between the calculated and experimental quantities need not be straightforward. Typically, the "observed" quantity results also from an assumed theoretical model, which need not be compatible with the assumptions involved in the ab initio calculations. (For a detailed analysis of the problem see Ref. 385). For example, the bond lengths are most usually determined from the dependence of the total energy on the positions of nuclei (in the Born-Oppenheimmer approximation, see Chapter 1.). The interatomic distance corresponding to the lowest energy is not compatible with the experimental bond length because the latter is not only due to the electronic energy but it is also affected by the vibrational motion. Hence, a rigorous comparison requires that a correction of the observed value for this effect be performed, i.e., the calculated equilibrium distances should be compared with the spectroscopic quantities r_e and not r_o. Corrections are also needed for rigorous treatments of inversion barriers, heats of formation, energies of activation and other quantities.

Another problem of a rigorous comparison of ab initio results with experiment is encountered with any observable which is determined by a polynomial fit to calculated points. Some molecular properties (mainly spectroscopic constants) depend on the fitting procedure rather strongly and if an inappropriate fit is used the discrepancies with experiment which are found may be erroneously assigned to basis set or correlation effects.

5.A. Molecular Geometries

The prediction of equilibrium geometries is one of the most topical tasks involved in chemical applications of the molecular orbital theory. The accuracy achieved depends of course on the quality of the wave function used. Some typical results are presented in Table 5.1.

T a b l e 5.1
Typical results for molecular geometries

Basis set	Correlation energy included by	Errors in predicted geometries	Representative references
STO-3G	-	$\sim 0.02 \times 10^{-10}$ m for AH bonds; $\sim 0.03 \times 10^{-10}$ m for AB bonds; $\sim 3-4^{\circ}$ for bond angles	386-389
4-31G, DZ and extended without polarization functions	-	$\sim 0.01 \times 10^{-10}$ m for both AH and AB bonds; overestimation of bond angles in structures as H_2O and NH_3	386,389,390
DZ+P	-	$\sim 0.01 \times 10^{-10}$ m; $1-2^{\circ}$; bond lengths mostly shorter than experiment	100,390
Hartree-Fock	-	for some diatomics bond lengths as much as 0.05 $\times 10^{-10}$ m too short	24,26
DZ+P	2nd order MB-RSPT or CI-SD	average error in AH bond lengths 0.003×10^{-10} m	253,390
DZ+P or larger	CEPA	error in bond lengths typically $< 0.005 \times 10^{-10}$ m and mostly in quantitative agreement with experiment	118,120,333

For practical purposes it is very important that even minimum basis sets such as e.g. STO-3G give very good molecular geometries. From comparative studies[386,391] it appears that STO-3G calculations provide molecular geometries that are more reliable than those given by popular semiempirical methods CNDO/2, INDO and MINDO/3. This is due to a fortuitous circumstance that basis set and correlation effects are mostly of opposite signs. As a consequence, the minimum basis set may sometimes provide bond lengths that match experiment surprisingly better than bond lengths given by near Hartree-Fock calculations that are underestimated in most cases (see Table 5.2). Our knowledge of

T a b l e   5.2

Effect of the correlation energy on the equilibrium bond lengths[a] (r_e in 10^{-10} m)

Molecule	near Hartree-Fock	PNO-CI	CEPA	Experiment
LiH	1.606	1.598	1.599	1.595
BeH	1.338	1.341	1.344	1.343
BH	1.221	1.235	1.238	1.232
CH	1.104	1.119	1.122	1.120
NH	1.018	1.034	1.039	1.037
OH	0.950	0.966	0.971	0.971
HF	0.898	0.912	0.917	0.917
NaH	1.916	1.891	1.891	1.887
MgH	1.725	1.724	1.728	1.730
AlH	1.647	1.644	1.645	1.646
SiH	1.516	1.523	1.526	1.520
PH	1.414	1.421	1.426	1.422
SH	1.331	1.339	1.344	1.341
HCl	1.266	1.273	1.278	1.275
NaLi[b]	3.000		2.873	2.826
N_2[c]	2.020	2.060	2.078	2.074
F_2[c]	2.525	2.606	2.666	2.68

[a] From Ref. 333 unless otherwise noted.
[b] Ref. 392.
[c] Ref. 120.

these trends derives predominantly from diatomics. Generally[385], extension of the basis set leads to lower r_e. Sometimes, however, r_e predicted even at the Hartree-Fock limit is larger than the experimental bond length, in which case both the basis set extension and the inclusion of the correlation energy lead to lower r_e.

In contrast to bond lengths, where good predictions are mostly obtained with minimum and DZ basis sets, accurate predictions of bond angles in certain structures are obtained only if polarization functions are included into the basis set. Consider for example the planar (D_{3h}) and pyramidal (C_{3v}) structures of ammonia. The highest occupied orbitals in two structures are of $1a_2''$ and $3a_1$ symmetry, respectively. It may be assumed (from perturbation theory) that the energy gain upon augmenting the basis set with d-functions is mainly due to their contributions to highest occupied orbitals. Hence, on symmetry grounds, d-functions centered on nitrogen can contribute significantly only to the pyramidal structure because none among the d-functions is of a_2'' symmetry. This explains[393] why the inversion barrier of NH_3 is well reproduced only with basis sets containing nitrogen d-functions. With a basis set including only s and p functions, the inversion barrier is too small and the equilibrium valence angle in the pyramidal structure is too large. If the bottom of the energy surface is shallow as in the case of H_3O^+, such a basis set may predict the planar structure to be lower in energy than the pyramidal structure[394].

Once the basis set and the type of the wave function is decided on, the next task is to perform an economic search for the energy minimum on the energy hypersurface. Since the geometry optimization is a very time consuming process, it is warranted to pay some attention to the optimization methods. There is a large variety of them. Garton and Sutcliffe[395] attempted to survey those of them, that proved useful or may become useful in quantum chemistry. A common feature of modern effective methods is that they require knowledge of the energy gradient and, profitably, also of the matrix of harmonic force constants or at least a reasonable estimate of the latter. In the forthcoming discussion, however, it should be kept in mind that the cost for the gradient evaluation is at least twice as high as the cost required for a single standard SCF run and that the evaluation of the force constant matrix is even more costly.

Suppose that the energy is a quadratic function of displacement coordinates, so that we may express it as

$$E = E_0 + \sum_i b_i q_i + \frac{1}{2} \sum_{ij} A_{ij} q_i q_j = E_0 + \vec{b}\vec{q} + \frac{1}{2} \vec{q}^+ \mathbf{A} \vec{q} \qquad (5.1)$$

Differentiation of (5.1) gives us

$$\vec{g} = \vec{b} + \mathbf{A}\vec{q} \qquad\qquad (5.2)$$

where $\vec{g}$ is the energy gradient with the components

$$g_i = \frac{\partial E}{\partial q_i} \qquad\qquad (5.3)$$

and where $\mathbf{A}$ is the matrix of harmonic force constants which is also referred to as the Hessian matrix. At the nearest stationary point $\vec{g} = 0$ and we may formally write

$$\vec{q}_o = -\vec{b}\,\mathbf{A}^{-1} \qquad\qquad (5.4)$$

so that for the direction vector $\vec{p} = \vec{q}_o - \vec{q}$, which extends from an arbitrary point $\vec{q}$ to the nearest stationary point, it holds

$$\vec{p} = -\mathbf{A}(\vec{q})^{-1}\vec{g}(\vec{q}) \qquad\qquad (5.5)$$

Assume, for the sake of simplicity, that the nearest stationary point is a minimum. In actual optimizations this should be established by the test that the eigenvalues of the Hessian matrix are all positive. Hence, once we know $\vec{g}$ and $\mathbf{A}^{-1}$ we should reach the minimum in just one step. Actual potential surfaces are, however, not strictly harmonic and eqn. (5.5) has to be applied in several successive steps. This is the essence of so-called Raphson-Newton or least-squares type algorithms. In order to avoid the calculation of $\mathbf{A}$, Pulay introduced[396, 397] the force relaxation method, in which a suitable initial guess to $\mathbf{A}$, $\mathbf{F}_o$, is kept unchanged throughout the whole iteration process

$$\vec{q}_{(i+1)} = \vec{q}_{(i)} - \mathbf{F}_o^{-1}\,\vec{g}_{(i)} \qquad\qquad (5.6)$$

The subscripts i+1 and i refer to two successive iterations. The choice of $\mathbf{F}_o$ does not affect the final geometry, but it only controls the rate of convergence. If use is made of "experimental harmonic" force constants, the optimum geometry is approached closely at the very first iteration[397], even if a diagonal F_0 is assumed. It may be anticipated[398], however, that the method would run into difficulties with complicated (e.g., non-classical) molecules, where $\vec{q}_0$ may be a poor approximation to the final geometry and where no realiable rules exist to estimate the F_0 matrix. In such cases it is preferable to compute the whole matrix of force constants using a small basis set or even a semiempirical MO method and to employ F_0 so determined in eqn. (5.6) in conjuction with the gradient evaluated by means of a more sophisticated approach.

From the above discussion it appears that efficiency of Newton-
-Raphson algorithms requires avoiding the expensive construction of A in each iteration. Instead, we try to replace the recalculation of A by some simple updating, as for example in the Marquardt-Levenberg method[395]. Updating A is a typical feature of so called quasi-Newton or variable metric methods, in which only the knowledge of $\vec{g}$ is required and the Hessian matrix is merely involved implicitly. The gen-
eral formula for this family of methods may be expressed as

$$\vec{q}_{(i+1)} = \vec{q}_{(i)} - \alpha \, H_{(i)} \vec{g}_{(i)} \tag{5.7}$$

Here α is a scale factor and $H_{(i)}$ is the estimate of A^{-1} at the i th iteration. Differences in the existing variants of the method are due to different ways of updating H. The most popular algorithms are due to Davidon, Fletcher and Powell[399], in which α is to be optimized in each search step, and Murtagh and Sargent[400], in which a fixed α may be chosen in a wide range. The property inherent to these algorithms is that they yield for quadratic surfaces the exact minimum and exact A matrix in N steps, N being the number of coordinates. Variable metric methods were introduced to quantum chemistry by McIver and Komor-
nicki[401] and they soon became popular in semiempirical calculations. Critical examination of the variable metric methods and the experience achieved in this field was reviewed by Pancíř[402]. In the original forms of the variable metric methods, the starting H matrix is chosen as the unit matrix. The fact that this preset of H brings about a rather slow convergence of the iteration process (5.7) is of little importance in semiempirical MO calculations, where the gradient evaluation is relatively easy[401,403]. In ab initio calculations, however, it is costly and some modification of the algorithm becomes necessary. Much faster convergence is already achieved[398,402] if H is preset with a diagonal matrix whose H_{ii} elements are of a suitable uniform value

or, preferably, if they are set equal to one of two different values, depending on whether the respective coordinate i is the bond length or bond angle. Evidently, the initial approximation to **H** is very important. Payne was able to show[404] that if a complete matrix of force constants is used, one iteration according to (5.7) gives the geometry prediction at the same accuracy level as the whole iteration process by means of the variable metric method.

For correct judging the efficiency of individual methods it is necessary to take into consideration the way in which the gradient and the force constant matrix are evaluated. There are two possibilities: either use is made of a direct analytical calculation[397] or the energy derivatives are obtained numerically by finite differences. The former way is more advantageous than the latter, the points in favor being the higher numerical accuracy attained and considerable savings in computer time and human effort. Until recently, analytical direct calculations were feasible only for first derivatives of the energy of SCF closed shell ground states. For possible extensions to second derivatives[405a], open shell configurations[405b], CI wave functions[405c], two configuration SCF[405b] and MC SCF wave functions[405d] we refer the reader to the cited literature[397,405]. Up to now several programs were developed[406-411]. With the most efficient among them, the calculation of the energy gradient requires approximately the same expense of the computer time as the complete SCF run. Pulay has shown[397] how efficient use of the analytical calculation of the energy gradient may be made in combination with the force relaxation method, though also the variable metric methods may work, if a reasonable guess to **H** is made.

An appealing feature of the numerical energy differentiation is that one may use any standard computer program and any type of the wave function. In this approach, the geometry optimization involves computing the total energy at many different geometrical conformations. Among various procedures reported we note two of them which keep the required number of data points to a minimum. Payne has shown[404] that the force constant matrix and the gradient may be evaluated using $1/2$ $N(N+3) + 1$ data points, the meaning of N being the number of geometry parameters to be optimized. As noted above, substituting $\vec{g}$ and **F** so determined into eqn. (5.6) gives us (in one iteration) the geometry at the same accuracy level as the variable metric method. In the approach suggested by Collins et al.[101] one first computes the gradient and the diagonal force constants, which are used for the construction of the (diagonal) $\mathbf{F_o}$ matrix. This requires 2N+1 data points. On applying eqn. (5.6) a vector $\vec{p}$ is determined (see eqn. (5.5)). Along

its direction a one-dimensional search is made using a quadratic procedure (3 more data points). In the next iteration F_0 is kept unchanged, so that only N additional points are necessary for the recalculation of $\vec{g}$ and 3 more points are necessary for a further one-dimensional search. Actually, the approach by Collins et al.[101] may be
viewed as a modified Davidon-Fletcher-Powell procedure. In most cases
it is sufficient to perform only two iterations, which means that altogether 3N+7 energy calculations are required. It should be emphasized, however, that the efficiency of the two methods depends on the
assumption that the energy is quadratic. This is satisfied in most
cases if the starting point is close to the optimum geometry.

Despite the recent progress, the geometry optimization of larger
molecules is still troublesome. Pulay et al.[411b] examined the possibility of extending the calculations to larger molecules with reasonable
cost by selecting a basis set especially suitable for the gradient e-
valuation (small number of primitives) which would be able to mimic
the results given by larger basis sets. They developed a new basis set,
denoted 4-21 (essentially $(7s3p/3s)/[3s2p/2s]$ with the shell structure), with which the gradient evaluation is about twice as fast as
with the 4-31G basis set. In spite of its modest size, the 4-21 basis
set proved very successful in describing molecular geometries. Another
approach to this problem was reported by Huber[412], who suggested the
use of basis sets, in which orbitals are located on N dummy nuclei instead of N atomic nuclei. This means that the nuclei and orbitals are
let to float independently and that in each iteration step of the
optimization process one computes 3N displacements for the dummies (orbital locations) due to the energy gradient and 3N displacement for
the nuclei due to the Hellmann-Feynman forces. The method is related
to the FSGO approach and its merit is the high flexibility of the basis set. As we learned in Section 2.G., small basis sets with floating
GTF's may be as effective as considerably larger standard basis sets
including polarization functions. With respect to standard optimization treatments, the profit from the higher flexibility of the basis
set overweights the additional cost due to the computation of Hellmann-Feynman forces and a somewhat poorer convergence of the optimization process. Huber showed[412] that time saving may also be a-
chieved if Hellmann-Feynman forces are used as an approximation to
the energy gradient in the first few iterations of a standard geometry
optimization. Although this way of using Hellmann-Feynman forces may
be expected to work well only with large basis sets and for molecules
having no atoms with polarized inner-shell electrons, the method is

worth of further development.

Finally, the transition states deserve a brief note. To prevent
the downhill movement along the reaction coordinate, McIver and Komornicki[413] proposed to minimize the norm of the gradient, $\sigma = \vec{g} \cdot \vec{g}$,
instead of the energy itself. Ab initio calculations of this type were
reported recently by Komornicki and collaborators[410]. A conceptually
simple approach was suggested by Halgren and Lipscomb[414] which is
based on the interpolation of internuclear distances by means of linear and quadratic transits from reactants to products. This interpolation is combined with optimization without the requirement of the
gradient evaluation.

5.B. Force Constants

Although force constants belong to the so called "spectroscopic
constants", they are not genuine observables but only values inferred
from the theoretical treatment of the assumed force field. It is usual
to express the force field in the form of a power series

$$V = V_0 + \sum_{i,j} \frac{1}{2!} f_{ij} q_i q_j + \sum_{i,j,k} \frac{1}{3!} f_{ijk} q_i q_j q_k$$

$$+ \sum_{i,j,k,\ell} \frac{1}{4!} f_{ijk\ell} q_i q_j q_k q_\ell + \cdots \tag{5.8}$$

where $q_1 \cdots q_n$ represent the displacement from the equilibrium geometry. Hereafter we shall assume internal coordinates so that q have
the meaning of Δr (in 10^{-10} m) or $\Delta \vartheta$ (in radians). The quadratic,
cubic, quartic, etc. force constants are defined as true derivatives
of V at the equilibrium

$$f_{ij} = \frac{\partial^2 V}{\partial q_i \partial q_j} \tag{5.9}$$

$$f_{ijk} = \frac{\partial^3 V}{\partial q_i \partial q_j \partial q_k} \tag{5.10}$$

$$f_{ijk\ell} = \frac{\partial^4 v}{\partial q_i \partial q_j \partial q_k \partial q_\ell} \tag{5.11}$$

The "experimental" force constants are determined in such a way that the true observables – the vibration energy levels – be fitted by the assumed potential (5.8).

The purpose of this introductory paragraph was to point out that for meaningful comparison of ab initio treatments with experiment the two force fields should be compatible. Ideally, one should calculate directly the vibrational energy levels. This is feasible for diatomic molecules either by solving the vibrational Schrödinger equation or by making use of a Dunham analysis, for example. The latter gives us the harmonic frequency, ω_e, and the anharmonicity constant, x_e, appearing in the expression

$$G(v) = \omega_e\left(v + \frac{1}{2}\right) - x_e\omega_e\left(v + \frac{1}{2}\right)^2 + y_e\omega_e\left(v + \frac{1}{2}\right)^3 - \ldots \tag{5.12}$$

to which the observed vibrational levels are fitted. A rigorous treatment of this type for polyatomic molecules is difficult and it is beyond the scope of this chapter. We shall therefore restrict ourselves to the harmonic approximation, though a series of accurate ab initio calculations have already been reported on higher order force constants of small polyatomic molecules. Hence, if the expansion (5.8) is truncated at the quadratic constants, one can arrive at the vibrational frequencies by means of Wilson FG matrix analysis. Although with polyatomic molecules the harmonic frequencies are only rarely available, the comparison with experiment is still meaningful – we shall see that deficiencies in the theoretical approach are sometimes much rougher than the difference between the harmonic and fundamental frequencies.

The actual values of force constants are determined by a polynomial fit to the computed energy points. It is rather disturbing that the calculated force constants may be strongly dependent[415] on the number and location of the energy values employed in a curve-fitting procedure. An instructive example was communicated by Pulay and Meyer[416], who recalculated the stretching force constant ($f_{rr} + 2f_{rr'}$) for NH_3 from a more reliable fit to the data points reported by Body et al.[393] They arrived at the value of 7.15 instead of 11.20 x 10^2 N/m.

This source of numerical inaccuracy is largely eliminated if the force method[396,397] is used. This method involves only a single numerical differentiation of the energy because the force constants are evaluated from forces on the nuclei, i.e., from the first derivatives calculated analytically. Another problem encountered with SCF calculations is the choice of the geometry for which the energy derivatives should be calculated. The choice of the calculated equilibrium geometry might appear to be the most natural. However Schwendeman[417] has shown that errors in the computed equilibrium geometry (with respect to experiment) may lead to first-order errors in the force constants. According to his suggestion it is preferable to use experimental geometries instead of calculated optimum geometries. If Schwendeman's procedure is followed, the basis set dependence of the calculated force constants is much smaller[418] and better agreement with experiment is achieved than with force constants calculated at the optimum geometry. A striking example was provided by Schlegel et al.[408] for NH_3 at the 4-31G level. The f_{rr} and $f_{\vartheta\vartheta}$ constants are 8.115×10^2 N/m and 0.467×10^{-18} Nm for the optimum 4-31G geometry, and 7.505×10^2 N/m and 0.748×10^{-18} Nm for the experimental geometry, compared to experimental values of 7.052×10^2 N/m and 0.636×10^{-18} Nm.

Several sources of uncertainties noted above are a probable reason why the trends indicated in Table 5.3 for the calculated force constants are not so clear-cut as with molecular geometries. What is meant by the "excellent agreement with experiment" in the last line of Table 5.3 in specified by the entries of Tables 2.17 and 5.4. We present here Table 5.4 for two purposes. First, it represents the present top calculations in this field and, second, it shows the effect of the correlation energy on the stretching vibrations.

To conclude this section in a practical way, let us comment on the possibilities of SCF calculations with medium basis sets. It is important that coupling force constants given by medium basis sets are accurate to about 0.1×10^2 N/m. Experimentally, the coupling force constants are often rather uncertain, so that even a correctly predicted sign may be valuable for spectroscopists in the construction of force fields for polyatomic molecules. The same may also be said about the cubic and diagonal quartic force constants. Hence one may arrive at reliable force fields in a very practical way by the combination of experimental diagonal quadratic and ab initio off-diagonal force constants. Since the overestimation of calculated diagonal force constants is systematic, it is also possible to arrive at reliable force fields by making use of scaling factors. Pulay and Meyer[139,420] suggested

T a b l e 5.3
Typical results for harmonic force constants

Basis set	Correlation energy included by	Comments on the calculated force constants	Representative references
STO-3G	-	Genarally poor results; stretching constants 20-30% overestimated; erratic behavior of the other constants	387, 407, 419
Medium- -sized	-	Diagonal constants 5-20% overestimated	407, 411b, 419-421
Near Hartree- -Fock	-	Slightly better than with medium-sized basis sets	416
Large	large scale CI, CEPA or MB-RSPT	Excellent agreement with ex- periment	120,167, 326b, 333, 422, 423

that calculated diagonal stretching constants be reduced by 10% and diagonal bending constants by 20%, leaving the off-diagonal constants unchanged. This correction should also absorb the effect of anharmonicity. Blom and Altona[424] used another approach which is suitable for confirming assignments of uncertain frequencies. Consider for example their treatment of the propane molecule. Five types of force constants were assumed,

    I : diagonal C-C stretching
    II : diagonal C-H stretching
    III: diagonal bending and rocking
    IV : torsion
    V : all off-diagonal

and for each type a fixed scale factor was adopted. The scale factors were adjusted to reproduce experimental vibrational frequencies. For propane the following factors I-V were found at the 4-31G level: 0.884; 0.888; 0.793; 0.875 and 0.814. These values are very similar to those determined for ethane and cyclopropane which suggests a possible transferability of scale factors among related molecules. The results of the treatment for propane are presented in Table 5.5. The average error in the 116 frequencies of propane and its deuterated

T a b l e 5.4

Effect of the correlation energy on the stretching vibration[a] (ω_e in cm^{-1})

Molecule	near Hartree-Fock	PNO-CI	CEPA	Experiment
LiH	1428.2	1408.6	1401.5	1405.7
BeH	2145.0	2086.7	2064.6	2060.8
BH	2484.6	2378.0	2352.1	2366.9
CH	3044.1	2883.5	2841.7	2858.5
NH	3546.5	3342.6	3269.3	3282.1
OH	4054.5	3833.3	3743.6	3739.9
HF	4476.2	4251.5	4169.3	4138.7
NaH	1184.2	1172.1	1172.3	1172.2
MgH	1583.6	1515.3	1492.3	1497.0
AlH	1730.8	1703.6	1691.7	1682.6
SiH	2126.2	2061.3	2034.7	2041.8
PH	2487.6	2408.6	2365.9	2380.0
SH	2839.9	2730.7	2676.4	2697.0
HCl	3141.1	3034.0	2977.2	2991.1
NaLi[b]	249.6		260.1	256.8
N$_2$[c]	2742	2525	2417	2359.6
F$_2$[c]	1247.4	1150	945	924

[a] From Ref. 333 unless otherwise noted.
[b] Ref. 392.
[c] Ref. 120.

analogs is 10.5 cm^{-1} or 0.73 per cent.

5.C. Barriers to Internal Rotation and Inversion

Internal rotation and inversion are processes that satisfy very well the two requirements for the conservation of correlation energy noted in Section 4.B. This implies that the use of SCF calculations should be sufficient. Indirectly, this assumption was established by a good agreement of SCF barriers with experiment. There is also direct evidence provided by the correlated wave functions of Ahlrichs and coworkers. For the barrier to rotation in ethane they obtained[120] with a large basis set the SCF barier of 12.7 kJ/mole. Inclusion of electron correlation by means of CEPA reduced the barrier to 12.4,

T a b l e 5.5

Observed and calculated wavenumbers for propane[424]

ω (cm^{-1})			ω (cm^{-1})		
observed	4-31G with 5 scale factors	Error (per cent)	observed	4-31G with 5 scale factors	Error (per cent)
2977	2989	-0.4	2968	2973	-0.2
(2962)[a]	2907		2887	2911	-0.8
2887	2913	-0.9	1464	1462	+0.1
1476	1478	-0.1	1378	1387	-0.6
1462	1465	-0.2	1338	1351	-1.0
1392	1381	+0.8	1054	1055	-0.1
1158	1160	-0.2	922	919	+0.4
869	863	+0.7	2973	2979	-0.2
369	367	+0.5	2968	2927	+1.4
2967	2963	+0.1	1472	1470	+0.2
1451	1453	-0.2	1192	1210	-1.5
1278	1282	-0.3	748	762	-2.0
(940)	893		268	255	+4.8
216	216	-0.2			

[a] Values in parentheses are uncertain and they were not used in the adjustment of scale factors.

very close to the experimental value of 12.3 kJ/mole. Nevertheless, the SCF barrier was improved only by 0.3 kJ/mole. A similar result was obtained[118] for the inversion barrier of NH_3. Electron correlation through CEPA raised the SCF barrier of 21.8 kJ/mole to 23.4 kJ/mole, compared to the experimental value of 24.3 kJ/mole. Hence the electron correlation may be safely disregarded. There are, however, two effects that are considerably more important: the presence of polarization functions in the basis set and the allowance for molecular relaxation in each conformation. The first effect and its origin were already noted with molecular geometries in Section 5.A. The importance of this effect may be seen from the reported calculations for NH_3. For example, with the 4-31G basis set, the inversion barrier is drastically under-estimated[88] (1.7 kJ/mole). Rauk et al.[97] used a considerably larger basis set (13s8p/8s2p)/[8s5p/4s2p] but arrived at a comparably low barrier of 5.0 kJ/mole. However, upon augmenting the basis set with

two d-sets on nitrogen, the barrier became 21.2 kJ/mole. The history
of the development of the theoretical approach to the NH_3 barrier de-
monstrates clearly how the use of an basis set of the insufficient
size may lead to incorrect conclusions about the role of the correlation energy. The correlation energy contribution obtained[425] for the
(4s2p/1s1p) STO basis set increased the barrier by 11.8 kJ/mole. In
contrast, with the [6s2p1d/3s1p] basis set the correlation energy
lowered[426] the barrier by 12.4 kJ/mole. In the first case the nitrogen
polarization functions were lacking, in the second case the number of
p-type functions was too low, though the basis set was rather large.
Calculations with more extended basis sets established[97,427] a very
good agreement at the SCF level. Direct calculations[118,428] on the
correlation energy with large basis sets proved that it is indeed very
small. Presence of the d-functions in the basis set is also important
for the internal rotation in H_2O_2. We comment, however, on this molecule for another reason, i.e. that it represents a case in which adequate results are obtained only if in addition to inclusion of d-functions also the full geometry optimization is performed for each conformation[429-431]. For a rigid rotation with the experimental $\angle OOH$,
R_{OH} and R_{OO} parameters, Dunning and Winter[430] arrived with the
[4s3p1d/2s1p] basis set at cis and trans barriers of 4251 and 94 cm^{-1},
respectively. If allowance was made for molecular relaxation, the barriers obtained were 2921 and 384 cm^{-1}, in excellent agreement with experimental values 2649 and 386 cm^{-1}.

To conclude this section it is possible to state that the origin
and magnitude of errors in the calculated barriers are, at least for
simple molecules, well understood. The knowledge accumulated permits
to decide whether a rigid rotation model and a basis set without d-
-functions are appropriate to a particular molecule. Valuable information on this topic was contributed by systematic 4-31G studies on
single and double rotors in small organic molecules which were reviewed by Pople[432]. A comprehensive review on barriers to rotation and
inversion was reported by Payne and Allen[433], which also contains almost complete bibliography of ab initio calculations.

5.D. Potential Curves

The applications treated in Sections 5.A.-5.C. refer to the por-
tions of energy hypersurfaces that represent the nearest surroundings
of equilibrium geometries. Here we comment on more remote regions of
the energy hypersurfaces in which there is a tendency to the electron

pair rupture. As we have learned in Chapter 4., the effect of electron
correlation is extraordinarily large in such a case. In this section
we shall inquire to what extent this effect is amenable to treatments
by the methods noted in Chapter 4. In Fig. 5.1 we present data for var-

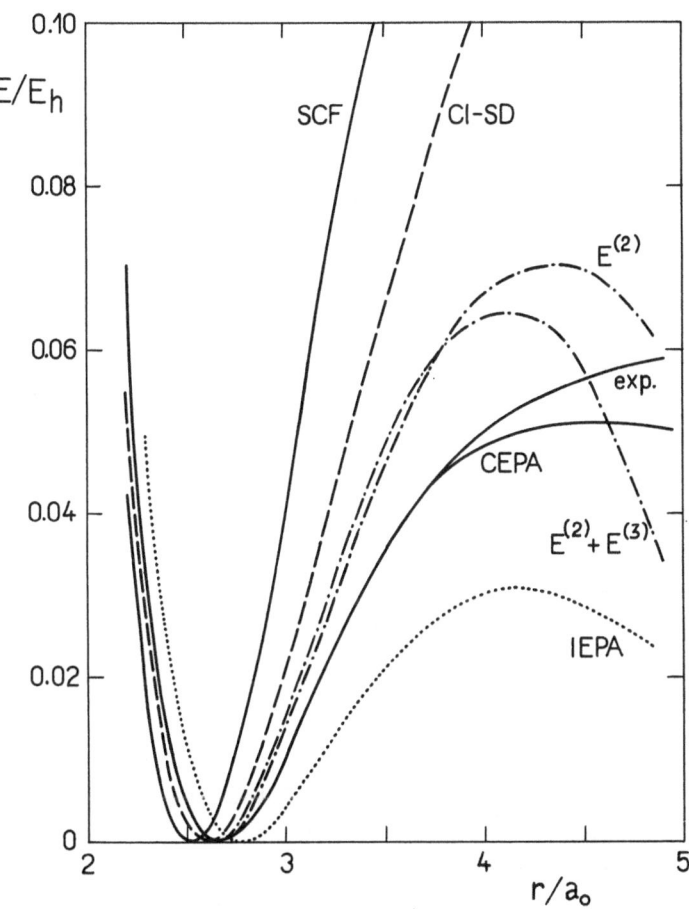

Figure 5.1
Potential curves of F_2 with different approximations[120,352].

ious levels of the approach to the potential curve of F_2. As we have
learned in Section 4.B., the Hartree-Fock approximation is especially
poor in this case, because it does not lead to a correct dissociation
limit, i.e. to the atoms in the Hartree-Fock ground states. In the
case of F_2, the proper dissociation is achieved by adding the excited

configuration $1\pi^4 3\sigma_u^2$ to the ground state $1\pi^4 3\sigma_g^2$. Generally, however, it is not sufficient to include just one configuration in the ground state wave function. For example, to ensure a correct dissociation, it is necessary to consider[255] 3 excited configurations for the $C_2 \rightarrow 2C(^3P)$ process and 9 excited configurations for the $N_2 \rightarrow 2N(^4S)$ process. This type of the conventional CI wave function was denoted by Lie and Clementi[255] as the "Hartree-Fock with proper dissociation" (HFPD). Although the HFPD functions give physically reasonable curves in the sense that at dissociation limits the energy is equal to the sum of the ground state SCF energies of the respective atoms, they still give poor dissociation energies. As seen in Fig. 5.1, CI-SD also fails to give a reasonable dissociation energy. The failure may be understood from the inspection of Table 5.6. It is seen that as the internuclear distance is increased, the coefficient of the $1\pi^4 3\sigma_u^2$

T a b l e 5.6

Coefficients (c^2) of $1\pi^4 3\sigma_g^2$ and $1\pi^4 3\sigma_u^2$ configurations in the PNO-CI wave function[120] of F_2 for different internuclear distances

r/a_o:	2.6	3.0	3.5	4.0
$(1\pi^4 3\sigma_g^2)$	0.920	0.900	0.844	0.765
$(1\pi^4 3\sigma_u^2)$	0.0237	0.0514	0.1111	0.1906

configuration increases and the function of the ground state $1\pi^4 3\sigma_g^2$ loses its meaning of the leading configuration. From this it follows that for a correct evaluation of the correlation energy it is not sufficient to assume double excitations only with respect to the configuration $1\pi^4 3\sigma_g^2$ but also with respect to the configuration $1\pi^4 3\sigma_u^2$. In the language of the traditional CI it is therefore necessary to assume also quadruply excited states that are double excitations with respect to the $1\pi^4 3\sigma_u^2$ configuration. Formulated generally[99], "a reasonable potential curve is obtained by brute force CI only if it includes all single and double excitations with respect to all configurations which enter the wave function with coefficients greater than, say, 0.3, at any internuclear separation". We can now understand that any method, which takes no account of quadruple excitations, must inevitably run into difficulties at larger FF internuclear distances. In Chapter 4. we learned that in no approach among those assumed in Fig. 5.1 the effect of quadruple excitations is included completely. CEPA includes those quadruple excitations that bring about cancellation of the renormalization term involved in the CI-SD approach and it gives, accordingly, a much better result than CI-SD. However, other quadruple excitations that are necessary for the proper dissociation are not included in CEPA and this is why CEPA tends to deviate from experiment at large distances. The same problem also arises at larger distances with the MB-RSPT through second and third orders. Here, however, an additional factor is involved, which follows from the very nature of the perturbation expansion. As the internuclear distance is increased, the perturbation (i.e. the correlation energy with respect to the one-determinant Hartree-Fock function) becomes larger and the convergence of the perturbation expansion becomes increasingly poorer. From this we can now understand why CEPA is far the best of the approaches used in Fig. 5.1. When speaking about the potential curve of the F_2 molecule, a paper by Das and Wahl[434] should be quoted. The results of their MC SCF OVC calculations match closely the experimental potential curve of F_2 in any region. These results document the high accuracy attainable by quantum chemical calculations, though the particular method requires much experimentation and its applicability to problems of chemical interest is rather limited.

5.E. Thermochemistry

The most typical problem involved in chemical applications of molecular orbital theories is the prediction of equilibrium geometries and relative energies of molecules. Its first part - geometry optimization - was dealt with in Section 5.A. Here we concentrate on the predictions of relative energies. Ideally, we would like to present a table for energies of reaction similar to Table 5.1 for molecular geometries. Since the trends are here not so clear-cut it is preferable to treat basis set and correlation effects separately. Basis set dependence indicated in Table 5.7 reflects the features of the plots of total energy versus the basis set size in Fig. 2.1 for H_2O and Fig. 2.2 for N_2. The energies for minimum basis sets are seen in the two figures to lie so high above the Hartree-Fock limit that it is hardly possible to expect such a cancellation of the basis set effect which would result in realiable predictions of energies of reaction. Predictions given by double zeta basis sets may be taken with confidence in treatments of

T a b l e 5.7

Errors in calculated energies of reaction with respect to Hartree-Fock data

Basis set	Accuracy achieved	Representative references
STO-3G and other minimum basis sets	unreliable predictions	
4-31G and DZ	within ∼ ±80 kJ/mole, poor results for molecules with multiple bonds and strain, mostly within ±40 kJ/mole if all reaction components are ordinary saturated molecules	117, 242, 258, 435
DZ+P	mostly within ±5-10 kJ/mole	242, 258, 435, 436
Valence-shell augmented with diffuse functions + polarization functions	suited for reactions involving negative ions or excited states if accuracy ±5-10 kJ/mole is to be achieved	121, 180

semiquantitative nature. It should, however, be kept in mind that energies of structures with multiple bonds and bent and pyramidal conformations are underestimated. The problem of conformations was already dealt with in Section 5.C. The extent of the defect of the DZ basis set with unsaturated molecules is illustrated by several hydrogenation reactions in Table 5.8. The DZ+P energies are used there as standards

T a b l e 5.8

Energies of reaction[123] (kJ/mole)

Reaction	DZ	DZ+P	Difference
$C_2H_2 + 3H_2 \rightarrow 2CH_4$	-502.8	-495.2	7.6
$H_2CO + 2H_2 \rightarrow CH_4 + H_2O$	-293.9	-254.5	39.4
$CO + 3H_2 \rightarrow CH_4 + H_2O$	-340.2	-262.8	77.4
$N_2 + 3H_2 \rightarrow 2NH_3$	-246.4	-171.3	75.1

as they are already close to the Hartree-Fock data. It would be of little use to compare the results in Table 5.8 with experiment because except for H_2 and H_2O the Hartree-Fock limits and correlation energies are uncertain. As a matter of fact, the exact nonrelativistic energies are even more hardly accessible than the Hartree-Fock limits and their estimated values are largely uncertain. Accordingly, the entries in Table 5.9 representing the correlation effects are indicatory rather

T a b l e 5.9
Correlation energy changes in chemical reactions

Type of reaction	Example	Correlation energy change (kJ/mole)	References
Both the number and neighborhood of electron pairs conserved	$H^+ + H^- \rightarrow H_2$	-2.5	241
	$H_2O + H^+ \rightarrow H_3O^+$	6.4	169
	$NH_3 + H^+ \rightarrow NH_4^+$	$\sim$13	436
	bond separation reactions	$\sim$12 (average)	437, 438
Only the number of electron pairs conserved	$HCN + 3H_2 \rightarrow CH_4 + NH_3$	$\sim$17	117, 243
	$CH_3CN \rightarrow CH_3NC$	17	370a
	$2BH_3 \rightarrow B_2H_6$	-66, -69	250, 370b
	$F_2 + H_2 \rightarrow 2HF$	$\sim$67	117, 241
Bond fission	$BeH \rightarrow Be + H$	-3	333
	$CH \rightarrow C + H$	96, 115	333, 436
	$FH \rightarrow F + H$	146, 176	333, 436
	$H_2O \rightarrow OH + H$	127, 152	253, 295, 436

than conclusive. We divided the reactions into three categories. The first one comprises processes that satisfy the two rules for the "conservation" of the correlation energy noted in Section 4.B. For these reactions the correlation energy change is small so that reliable energy predictions may be obtained at the SCF level. For the reactions of the second group only the first rule is satisfied: the number of electron pairs is maintained, i.e., all reaction components are closed shell molecules but the nearest environment of the pairs is changed. Among the examples given in Table 5.9, the correlation energy change in the reaction $F_2 + H_2 \rightarrow 2HF$ is extraordinarily high. The examples of reactions $HCN + 3H_2 \rightarrow CH_4 + NH_3$ and $CH_3CN \rightarrow CH_3NC$ are probably more typical. The reactions in the third group violate both rules. Intentionally, we included also in this group the dissociation of BeH which may appear at first sight to be a characteristic reaction of this type. Actually, this process belongs rather to the second group of our classification (Section 4.B.).

We now comment briefly on the conversion of calculated energy differences to observable quantities. The calculated energies of reaction, ΔE, correspond to energy differences between the bottoms of the potential surfaces of reaction components. To convert ΔE to the heat of reaction at absolute zero, ΔH_0^o, it is necessary to correct the calculated energies for zero-point energies. This is simply done by adding the term $1/2 \, hc \sum \omega_i$ to the energy of each reaction component, where the factor $1/2 \, hc$ is $5.98133 \, J.cm.mol^{-1}$. The wavenumbers of vibrational modes, ω, can be obtained from the calculated quadratic force constants by means of the Wilson FG matrix analysis. Usually, however, the experimental heats of reaction are known only for the room (or higher) temperature, T, and it is therefore necessary to transform ΔH_0^o to ΔH_T^o by means of the statistical thermodynamic treatment. Kosloff et al.[439] constructed very accurate partition functions for molecular hydrogen from ab initio data and arrived at the thermodynamic functions that differ from those based on spectroscopic data only in the third decimal place. However, for chemical purposes much simpler partition functions may be used; mostly the rigid rotor and harmonic oscillator approximation is quite sufficient.

5.F. Chemical Reactivity

A nonempirical approach to the chemical reactivity may of course be made along the same lines as has been practised for years in treatments by semiempirical all-valence electron methods. Typically, the results of such treatments provide qualitative explanation of the observed facts and give guidance for further experiments. Here we shall deal only with what may be taken as the ultimate goal of ab initio calculations in the field of chemical reactivity - the predictions of absolute values of equilibrium and rate constants.

In terms of statistical thermodynamics and the theory of absolute reaction rates[440], the equilibrium and rate constants of gas phase reactions

$$A + B \rightleftharpoons C + D \; ; \qquad K = \frac{p_C p_D}{p_A p_B} \qquad\qquad (5.13)$$

$$M + N \rightleftharpoons M\cdots N^{\ddagger} \longrightarrow \text{products} \quad ; \quad r = kp_M p_N \tag{5.14}$$

are expressed as

$$K = \frac{Q_C Q_D}{Q_A Q_B} e^{-\Delta H_0^o/RT} \tag{5.15}$$

and

$$k = \Gamma \frac{kT}{h} \frac{Q^{\ddagger}}{Q_M Q_N} e^{-\Delta H_0^{\ddagger}/RT} \tag{5.16}$$

To arrive at K and k, our task is to express the following terms appearing in eqns. (5.15) and (5.16): the partition functions (Q) of reactants, products and of the activated complex, the heat of reaction at absolute zero, ΔH_0^o, the enthalpy of activation at absolute zero, $\Delta H_0^{\ddagger}$, and the tunnelling correction factor, Γ. For an ideal gas the total partition function can be expressed within the rigid-rotor and harmonic oscillator (RRHO) approximation as a product

$$Q = Q_{tr} Q_{rot} Q_{vib} Q_{el} \tag{5.17}$$

where the individual components have the following forms

$$Q_{tr} = \frac{(2\pi mkT)^{3/2}}{h^3} \frac{RT}{p} \tag{5.18}$$

$$Q_{rot} = \frac{1}{\sigma} \left[\frac{8\pi^2 IkT}{h^2} + \frac{1}{3} + \frac{1}{15} \frac{h^2}{8\pi^2 IkT} + \frac{4}{315} \left(\frac{h^2}{8\pi^2 IkT} \right)^2 \right] \tag{5.19}$$

for linear molecules

$$Q_{rot} = \frac{(\pi I_A I_B I_C)^{1/2}}{\sigma} \left(\frac{8\pi^2 kT}{h^2} \right)^{3/2} \tag{5.20}$$

for nonlinear molecules

$$Q_{vib} = \prod_i \frac{1}{1 - e^{-h\nu_i/kT}} \tag{5.21}$$

For most molecules it is reasonable to set the electronic partition function to the statistical weight of the ground electronic state.

We now comment in general how to arrive from ab initio calculations at K and k. The whole procedure, outlined schematically in Fig. 5.2,

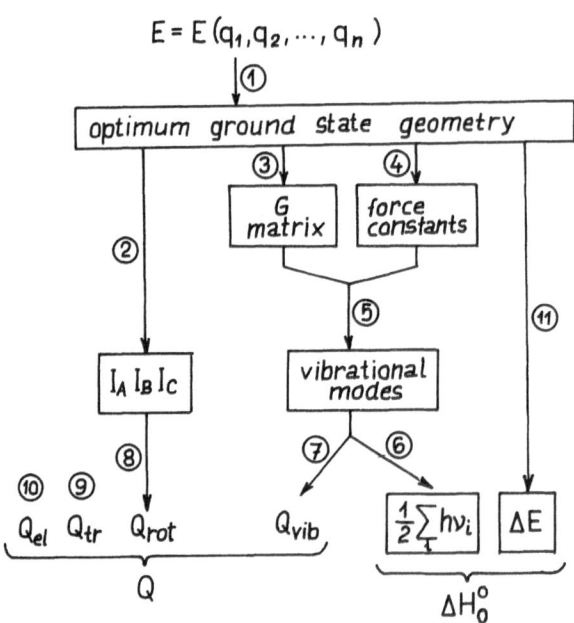

Figure 5.2
Computational scheme for the evaluation of partition functions and enthalpies at absolute zero.

is divided into 11 steps. The first step involves the determination of the ground state geometry. As we learned in Section 5.A., it is sufficient to perform this in most cases at the SCF level by making use of a small or a medium-sized basis set. A very accurate determination of the molecular geometry is not warranted because the errors in the molecular geometry affect K and k (through moments of inertia and the G matrix) rather little. The geometry optimization is followed by the evaluation of moments of inertia (step 2) and the G-matrix elements (step 3) according to simple formulas[441,442]. The next step (4), the

evaluation of force constants, was dealt with in Section 5.B. Also the
force constants may be calculated at the SCF level with a small or
medium-sized basis set. If the diagonal force constants are reduced by
standard scaling factors (see Section 5.B.), a very close agreement
with experiment may be expected. The errors in the computed vibrational frequencies affect the estimated equilibrium constant through
the vibrational partition functions and the zero-point energies. Whereas the former affect the equilibrium constant at lower temperatures
very little, the effect of the latter may be larger. When the matrix
of force constants is available, it is converted to vibrational frequencies by means of the FG matrix analysis[442] (step 5). From the vibrational frequencies one arrives easily at zero-point energies and
vibrational partition functions (steps 6 and 7). Also the evaluation
of the other components of partition functions is straightforward
(steps 8-10). The last step to be performed is the evaluation of the
heat of reaction at absolute zero, ΔH_0^o. Since the evaluation of ΔH_0^o is
mostly a crucial point for a successful estimation of the equilibrium
constant, the molecular energies must be calculated as accurately as
possible. Therefore a large basis set should be used and electron correlation should be accounted for. Fortunately, it is tolerable to perform this accurate but time-consuming calculation on the molecular energy (step 11) only once for the geometry which is optimum for a smaller basis set. It is believed that the error so introduced by using
slightly different geometries is unimportant. By correcting the computed molecular energies for zero-point energies one arrives at ΔH_0^o.

Probably the first study along these lines was reported by Clementi and Gayles[251], the famous prediction of the gaseous NH_4Cl. In a
later paper[443] of this type, the existence of the dimer $(LiH)_2$ was
predicted, for example. If the quantitative predictions of equilibrium
constants are to be competitive to experiment, the computed heats of
reaction should be accurate to within 4 kJ/mol, which implies that the
computed equilibrium constants differ from the observed ones by a factor of at most 5 at room temperature. Direct calculations of equilibrium constants show that this accuracy level, referred to as the
"chemical accuracy", is attainable[180,350,435] with reactions involving
small molecules.

With rate processes the situation is less clear-cut. In principle,
ab initio calculations permit a more sophisticated approach to rate
processes than that based on the transition state theory (TST) and
represented by eqns. (5.14) and (5.16)-(5.21). One may calculate the
whole energy hypersurface and apply to it molecular trajectory calcu-

lations. Treatments of this type are of primary importance for the
understanding of rate processes. Nevertheless it appears that only
TST will remain, for a long time to come, manageable for reactions
of real chemical interest. Although the underlying theoretical basis
of TST may be disputed (for a review see e.g. Ref. 444) and although
it is advocated to refer TST as to the "approximate" or even "empirical" rather than "absolute" theory of rate processes, a detailed a-
nalysis showed[445] that TST is capable of giving rate constants with
the accuracy comparable to experiment, provided that the treatment is
based on data given by highly accurate ab initio calculations. To demonstrate such a treatment, we selected the example[446] of the CH_4 +
H $\longrightarrow$ CH_3 + H_2 reaction. The data used for the evaluation of the rate
constant (5.16) are collected in Table 5.10. The tunnelling correction

T a b l e 5.10
Data used for the treatment of the process CH_4 + H $\longrightarrow$ CH_3 + H_2

Quantity	Its value	Origin
CH bond length in CH_4	1.093×10^{-10} m	a
Geometry of the activated complex $H_3CH^1H^2$: $r_{H^1H^2}$	0.903×10^{-10} m	b
r_{CH^1}	1.376×10^{-10} m	b
r_{CH}	1.095×10^{-10} m	c
$\sphericalangle$ HCH	114.7°	c
Zero point energy for CH_4	123.1 kJ/mol	d
Zero point energy for CH_5	112.4 kJ/mol	e
Classical barrier height, E_c	67.3 kJ/mol	f
Enthalpy of activation, $\Delta H_0^{\neq}$	56.7 kJ/mol	g
Imaginary wavenumber for CH_5	1400 cm^{-1}	h
Imaginary wavenumber for Eckart's tunnelling	1285 cm^{-1}	i

a Ref. 447.
b From UHF-CI-SD ab initio calculations[448].
c Assumed[448].
d Wavenumbers from the FG matrix analysis; force constants from 4-31G
  calculations[407].
e Wavenumbers from the FG matrix analysis; force constants from 4-31G
  calculations[449]
f Ref. 448.
g From E_c and zero-point energies.
h As in the footnote e but the stretching force constants obtained
  from a quadratic fit to energy values by UHF-CI ab initio calculations[448].
i Eqn. (5.22).

factor was calculated according to Eckart's approach[450] to the one-dimensional quantum mechanical tunnelling but with the modifications advocated by Jakubetz[451] for one-dimensional reactions. In our case this means that the height of the Eckart's barrier was set equal to $\Delta H_0^{\ddagger}$ and the imaginary frequency (corresponding to the vibrational mode with the negative eigenvalue and determining the width of the Eckart's barrier) was corrected as follows

$$h\nu_E = h\nu_S \sqrt{\frac{\Delta H_0^{\ddagger}}{E_c}} \qquad (5.22)$$

where ν_S is the imaginary frequency obtained directly from the vibrational FG analysis of the transition state and E_c is the "classical" energy of activation, i.e., the energy barrier given directly by ab initio calculations. The computed rate constants are presented in Fig. 5.3. It is seen that they are in a reasonable agreement with experiment, though the computational techniques used may be considered routine. If use is made of a fit[454] to experimental data, $k = 6.25 \times 10^{13} \exp(-48600\ J\ mol^{-1}/RT)$, the TST prediction for the rate constant at 298 K is underestimated by a factor of about 30 whereas if the tunnelling correction is applied, the factor reduces to ~ 5. The remaining discrepancy may be ascribed to the underestimated activation barrier [448], the error being 4-6 kJ/mol.

Utility of ab initio calculations in this field may be demonstrated with the reaction $H^* + FH \rightarrow H^*F + H$. Until recently, this reaction was believed to possess a low activation energy. A low activation barrier was also predicted by most semiempirical (BEBO and LEPS) treatments, the typical values being from 4 to 30 kJ/mol. In contrast, the ab initio calculations[455-457] predicted the activation barrier to fall in the range 180-205 kJ/mol. On the basis of experience accumulated with ab initio calculations it was possible to claim that the true barrier cannot be less than, say, 165 kJ/mol. In this way the ab initio calculations were able to give a realistic account of the $H^* + FH \rightarrow H^*F + H$ reaction prior to experimental confirmation[458].

5.G. Ionization Potentials

Calculations of ionization potentials represent a valuable tool for the interpretation of electron spectra. Unlike in semiempirical

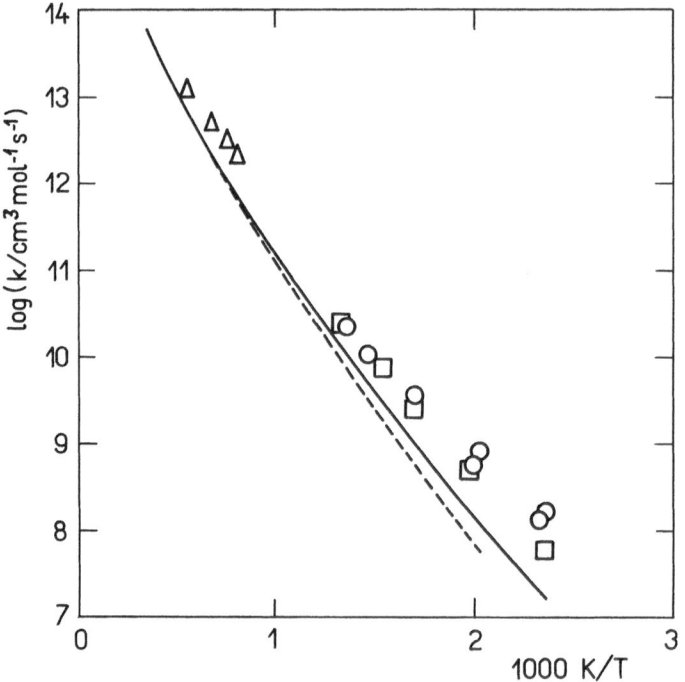

Figure 5.3

Temperature dependence of the rate constant for the process $CH_4 + H \rightarrow CH_3 + H_2$. The dashed and full lines, respectively, represent the TST treatments without and with applying Eckart's correction for the quantum mechanical tunnelling. Experimental data are taken from Ref. 452 ($\triangle$), Ref. 453 ($\square$) and Ref. 454 ($\bigcirc$).

calculations, in ab initio calculations all electrons are treated explicitly so that it is not necessary to make any distinction between the ionizations of valence shell and inner shell electrons. For this reason we do not distinguish between the Photoelectron Spectroscopy (PES) and the Electron Spectroscopy for Chemical Analysis (ESCA), though the theoretical approach to the latter involves some specific problems.

The simplest theoretical approach to ionization potentials is based on the Koopmans' theorem[459] which relates the h-th ionization potential to the negative value of the Hartree-Fock orbital energy, ε_h, of the parent closed shell system

$$IP_h^K = -\varepsilon_h \tag{5.23}$$

According to the Koopmans' theorem the photoelectron spectrum should exhibit as many bands as there are occupied molecular orbitals. Roughly speaking, this is actually observed in most cases so that the orbital structure of atoms and molecules is proved in this way by a direct experimental evidence[460]. Quantitatively, however, eqn. (5.23) does not provide a satisfactory account of the observed spectra. This is due to three problems[461] inherent in the use of Koopmans' theorem:

(i) A self-consistent adjustment of the orbitals of the ion is disregarded, i.e., all occupied orbitals are assumed to be unaltered when going from molecule to ion. Accordingly, the energy, ΔR, associated with this effect is called reorientation, reorganization or relaxation energy.

(ii) The change in the correlation energy, ΔC, on going from molecule to ion is not accounted for.

(iii) It is assumed that the relativistic energy is the same in both molecule and ion.

We comment first on the last point. As we have learned in Chapter 1, the relativistic energies are large for inner shells and their values increase with the atomic number. This trend is reflected in the relativistic corrections. For example, for the inner-shell ionization potentials the relativistic corrections are 0.1 eV[462a] for CH_4, 0.8 eV[462b] for Ne, and 14 eV[463] for Ar. Since the parent system is richer by one electron than the corresponding ion, the relativistic energy in the former is larger than it is in the latter. This means that upon correcting the ionization potentials for relativistic effects one arrives at higher values. Since the relativistic corrections are not accessible to direct calculations, we are forced to adopt the following two assumptions. First, the relativistic effects may be ignored with ionizations of outer-shell electrons and, second, the ionization potentials predicted by ab initio calculations for inner shells may be considerably underestimated owing to the neglect of relativistic effects.

Hence disregarding the relativistic effects gives us

$$IP_h^{KRC} = -\epsilon_h + \Delta R + \Delta C \tag{5.24}$$

The difference $IP_h^{KRC} - IP_h^K$ is called the Koopmans' defect. The reorganization energy, ΔR, is accountable for within the SCF approximation by expressing the ionization potential as

$$IP_h^{KR} = E_h^{SCF} - E^{SCF} \qquad (5.25)$$

where E^{SCF} is the total SCF energy for the parent molecule and E_h^{SCF} the total SCF energy for the respective ionized (hole) state. This means that evaluation of each ionization potential requires an additional SCF calculation. The procedure is referred to as the ΔSCF method and ΔR for any ionization potential is given by the difference $IP_h^{KR} - IP_h^K$. Unlike Koopmans' theorem, the ΔSCF method gives us a possibility to obtain a theoretical prediction for the adiabatic ionization potentials by performing the SCF calculations of ion at the geometries optimum for its particular electronic states. Computationally, however, it is simpler to evaluate the vertical ionization potentials, i.e., to use the same geometry as for the parent system, in which case the integrals over basis set functions obtained for the parent system need not be recalculated in open shell calculations.

In the applications of the ΔSCF method to the 1s hole states, the SCF calculation is performed for an electron configuration with the singly occupied 1s orbital. This is not justifiable on theoretical grounds because there may be several states of the same symmetry that are lower in energy (for example the $2a_1$ hole state of methane). As a consequence the wave function for the 1s hole state obtained in this way does not yield an upper bound to the true energy. Fortunately, for practical purposes the error so introduced is unimportant[463] owing to a large energy separation between the core hole state and lower states of the same symmetry. The ΔSCF approach to the core electron ionization has a remarkable feature that it is preferable to impose no symmetry restrictions on the molecular orbitals of symmetrical molecules. This is the case of molecules with several equivalent atoms whose 1s orbitals may be combined into delocalized core orbitals. To be more specific, consider the case of O_2. Here the combination of the two 1s orbitals gives $1\sigma_g$ and $1\sigma_u$ delocalized orbitals. Electron ejection from either of them brings about a delocalized core hole state. Consider now another possibility viz. the formation of a localized hole state by ejecting the electron from one of the two 1s orbitals. According to calculations on O_2^+ by Bagus and Schaefer[464] the localized hole state is favored by 12 eV over the delocalized state and the ionization potential for the former is in much better agreement with experiment than it is for the latter. A detailed theoretical analysis of the problem was given recently by Cederbaum and Domcke[465] who showed that if delocalized orbitals are used, both reorganization and cor-

relation effects are important and of the same order of magnitude.
When localized orbitals are used, the correlation effects are much
smaller than the reorganization effects. Since the ΔSCF method accounts only for the reorganization effects, better agreement with experiment is obtained with the localized orbitals.

Koopmans' theorem is also expressible as a difference between the
total energies of ion and molecule

$$IP_h^K = E_h^{SCF} \text{ (frozen)} - E^{SCF} \qquad (5.26)$$

but in contrast to eqn. (5.25) the total energy of the ion is obtained
within the frozen orbital approximation, i.e., the wave function for
the ion is assumed to be built up from the molecular orbitals of the
parent system. A direct SCF calculation on the ion gives of course a
lower energy so that the ΔR energy is negative (see Fig. 5.4). On the

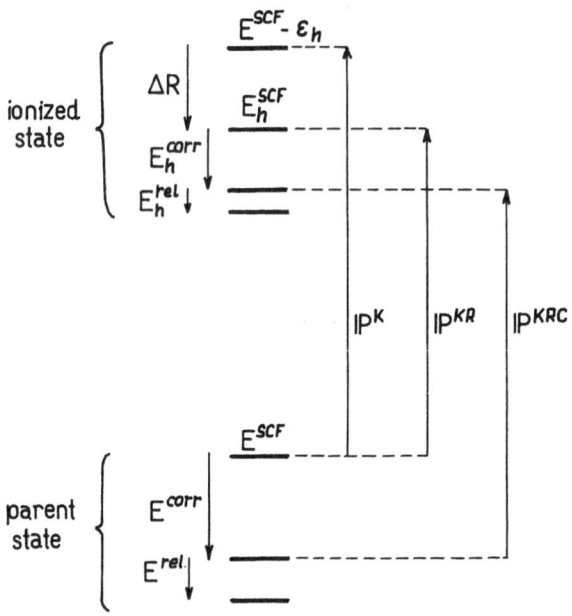

Figure 5.4
Schematical representation of corrections
to the Koopmans' theorem

contrary, the correlation energy correction, ΔC, may be expected to be mostly of positive sign because, compared to an ionized state, the parent molecule is likely to have more correlation energy. Indeed, ΔR and ΔC cancel usually to certain extent which affords the explanation why the Koopmans' theorem sometimes works well. Generally, however, the cancellation is only approximate and in some exceptional cases ΔC may be even negative, so that ΔR and ΔC are of the same sign. The importance of rearrangement and correlation energies may be judged from the entries of Table 5.11 which presents the results obtained by Meyer

T a b l e 5.11

Ionization potentials[a] of H_2O and CH_4

Hole state	Koopmans' theorem	Δ SCF	PNO-CI	Experiment[b]
H_2O				
$1b_1$	13.86	11.10	12.34	12.61 (12.78)
$3a_1$	15.87	13.32	14.54	14.73 (14.83 $\pm$ 0.11)
$1b_2$	19.50	17.59	18.73	18.55 (18.72 $\pm$ 0.22)
$2a_1$	36.77	34.22	32.25	32.2
$1a_1$	559.48	539.11	539.53	539.7
CH_4				
$1f_2$	14.85	13.67	14.23	14.4 (14.6)
$2a_1$	25.69	24.31	23.67	23.0 (23.2)
$1a_1$	304.86	290.76	290.70	290.7 (290.9)

[a] Taken from Refs. 292, 329; all entries in eV.
[b] Values corrected for vibrational effects in parentheses.

for H_2O[292] and CH_4[329] with large basis sets. The energies for the two molecules are the best variational energies attained so far. From the differences between the entries in the first and second columns it is seen that ΔR is significant particularly for inner-shell ionization potentials, where its recognition leads to a good agreement with experiment. Typically, Koopmans' theorem gives too high ionization potentials whereas the ΔSCF approach gives too low ionization potentials (except for the $2a_1$ hole states of the two molecules).

The entries of Table 5.11 demonstrate the fact that by making use of the Koopmans' theorem we can interpret the electron spectra only

qualitatively. In some cases the Koopmans' defects are so large that
not even the order of the observed states is correctly reproduced.
There are many examples of such a "breakdown of the Koopmans' theorem". We selected among them the case of ferrocene. Several of its
lowest ionization potentials are presented in Table 5.12. Obviously,

T a b l e   5.12

Ionization potentials (in eV) of ferrocene[466]

Hole state	Koopmans' theorem	ΔSCF	Experiment
e_{2g}	14.4	8.3	6.8
a_{1g}	16.6	10.1	7.2
e_{1u}	11.7	11.1	8.8
e_{1g}	11.9	11.2	9.3
a_{2u}	16.0	15.5	-

Koopmans' theorem cannot be applied to this molecule. The ΔSCF values
are still in a rather poor agreement with experiment (partly owing to
a small basis set used), but the order of ionization potentials is
correct. The breakdown of the Koopmans' theorem may be elucidated[466]
by a different extent of the electron reorganization involved in electron ionizations from the respective molecular orbitals. For ligand
orbitals e_{1u}, e_{1g} or a_{2u}, there is little rearrangement upon ionization and the ΔSCF values differ very little from the corresponding
orbital energies (see Table 5.12). For a metal orbital (a_{1g} or e_{2g})
there is a marked rearrangement upon ionization: while these orbitals
include a small amount of ligand orbitals for the parent molecule,
they become nearly pure metal orbitals for the ion. In general, the
breakdown of the Koopmans' theorem occurs very often for ionizations
from the 3d orbital of transition metal complexes (several examples
are cited in Ref. 467).

The breakdown of the Koopmans' theorem with the nitrogen molecule
[468,469] is notable because of its basis set dependence[59]; with the DZ
basis set the order of orbital energies agrees with experiment whereas with the [4s3p] and larger basis sets the breakdown of Koopmans'
theorem occurs. Incorrect order of the $2\sigma_u$, $3\sigma_g$ and $1\pi_u$ ionization
potentials is predicted even by the near Hartree-Fock ΔSCF calculations[24]. This suggests that the correlation effects are extraordinar-

illy important for this molecule[468,469].

A systematic study of the effect of bond lengths in ionized systems (and also of the external electric field) on the reorganization energy was communicated by Clark and co-workers[470]. The experience accumulated established the utility of the ΔSCF method as the first step in improving the results given by the Koopmans' theorem. It is evident, however, that the correlation effects cannot be disregarded if a highly accurate treatment is attempted. In principle, in order to include correlation energy use may be made of any method noted in Chapter 4. It depends of course on the level of sophistication of the method used. On the one extreme, with the simple techniques such as the EPCE-F2σ method, it is hardly possible to attain[269] better accuracy than ±0.5 eV. The other extreme may be represented by the PNO-CI data presented in Table 5.11. Standard configuration interaction is usually not used with the aim of including correlation energy. It is used rather for giving an account on the so-called satellite peaks (vide infra) or to make allowance for orbital relaxation[471] by means of the configuration interaction with singly excited doublet states, if the wave function for the leading open shell configuration is constructed from the closed shell orbitals. The variety of approaches used to account for both relaxation and correlation effects is very large: PNO-CI and CEPA[334], techniques based on Green's functions[468, 472-476], diagrammatical MB-RSPT[320,477,478], the equation of motion method[479,480], the natural transition orbitals method[481], the ordinary RSPT[104,482], the superoperator technique[476,483], and the method of density matrices and natural functions[484]. Among these methods the Green's function approach seems to be the most elegant and effective one. It is based on the fact that the one-particle Green's function, G, gives exact ionization potential in a given basis set. Feasibility of the calculations is due to the circumstance that G is expressible through so called self-energy part by means of the diagrammatic expansion in terms of interaction matrix elements (the latter are actually the transformed two-electron integrals over MO's). The expansion of the self-energy part contains all second and third order diagrams whereas the higher order terms are approximated by an infinite summation of ring and ladder diagrams. Explicit expressions for the ionization potentials in Green's function approach may be found in Ref. 472. It is an advantage of the method that only the wave function for the electronic ground state is required. Correlation and reorganization effects are explicitly taken into account, though the ionization potentials are not calculated as a difference in energy between the hole state and the parent molecule. The method was applied to a variety of both small and large molecules. As an example we selected in Fig. 5.5 the treatment of the pyridine molecule[486]. It is seen from

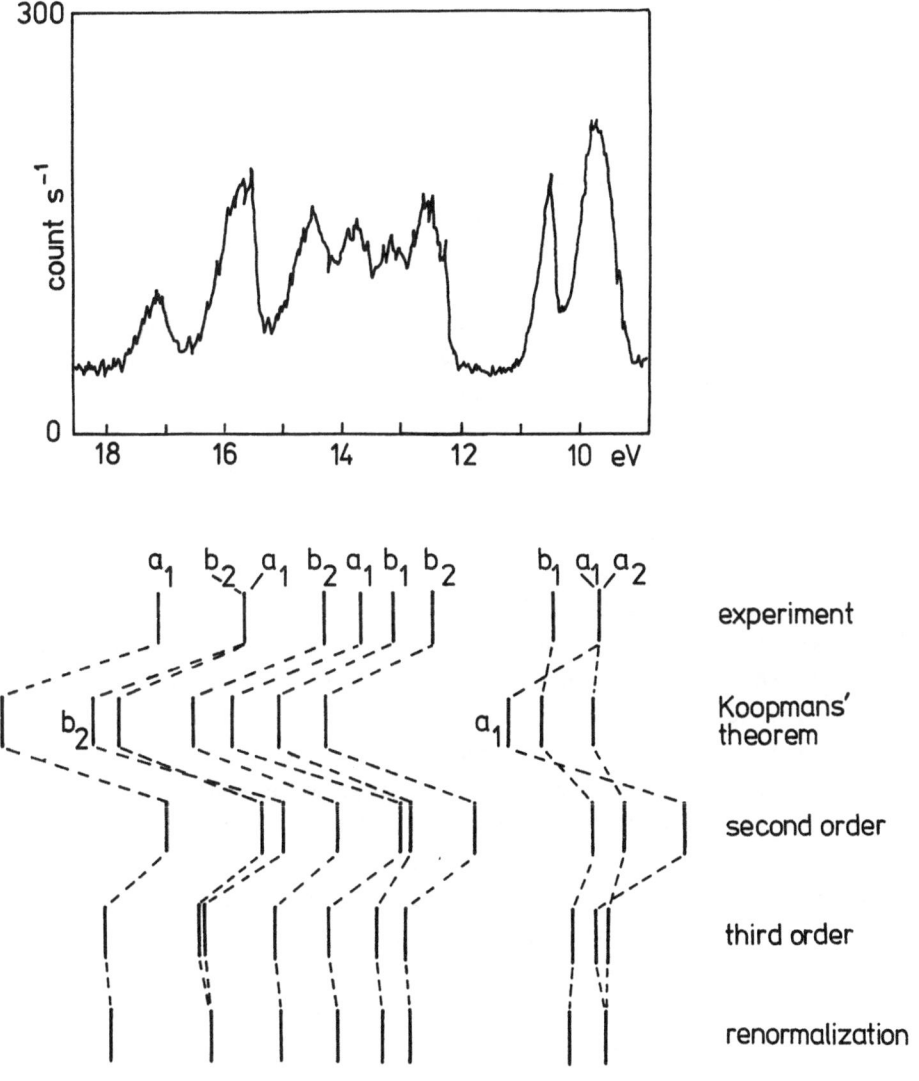

Figure 5.5

He (I) photoelectron spectrum of pyridine[485] and the vertical ionization potentials calculated[486] with the DZ basis set by means of Green's functions.

Fig. 5.5 that again the Koopmans' theorem does not hold. The second order of the perturbation expansion reproduces correctly the order of ionization potentials. However, the values are too low, because the second order corrections to the Koopmans' theorem are greatly overestimated. This effect is a general property of many-body methods and it was found not only with the Green's functions technique but also with the MB-RSPT calculations[478]. For small molecules the defect of the second order is even larger, so that the predicted ionization potentials are in some cases inferior to values given by the Koopmans' theorem[472]. For the pyridine molecule[486], the second order predicts an appreciable energy gap between $3b_2$ and $5a_1$ ionization potentials, in disagreement with experiment. The same situation is met with the $7a_1$ and $1a_2$ ionization potentials. To summarize, the use of the second order in the many-body theory is insufficient. As it is seen in Fig. 5.5 for the pyridine molecule, making use of the third order removes all apparent discrepancies, whereas the further improvement by means of the renormalization brings about only a slightly better agreement with experiment. From the practical point of view, the calculations provided an important finding viz. that disregarding several of the highest virtual molecular orbitals affected the results very little. This effect is small compared to errors in the orbital energies given by the DZ basis set.

Closely related to the Green's function method is the approach based on MB-RSPT[320,477,478]. The two methods yield essentially equivalent results. The computational procedure in MB-RSPT is similar to that for the ground state correlation energy of closed shell systems (see Chapter 4.). The number of diagrams is somewhat larger, however, and in addition to two-electron integrals over MO's that are needed in the ground state calculations up to the third order, in calculations of ionization potentials we need also integrals of the (oo|ov) and (ov|vv) types. The transformed molecular integrals for the ionized system are not required. Hence, as with the Green's function approach, the MB-RSPT calculations are profitable that the ionization potentials are calculated directly from the wave function of the parent system. This is enabled by a formal cancellation of those terms that contribute equally, at least up to the third order of the perturbation theory, to the correlation corrections of both the ionized state and the parent state. The remaining terms can be represented diagrammatically.

The cancellation noted above with the many-body approaches has two important consequences. First, no error is introduced into the result by subtracting the energies of the two systems. The second point in favor, which is perhaps more important than the first one, is a lower numerical effort. The first point refers mainly to CI calculations. An interesting comparison referring to this problem was reported[487] for the ionization potentials of ethylene computed by the Green's function method and the CI method[278,286] with the Langhoff and Davidson's correction (see Chapter 4.). The results given by the two approaches were in quite good agreement. The authors of the quoted paper [487] also examined the basis set effect (in particular the effect of polarization functions) on the correlation effects in both the ground and several ionized states of C_2H_4.

It should be noted that ab initio calculations may also be applied to fine effects observed in electron spectra such as the vibrational structure of bands[472,488-493], shake-up and shake-off processes[460, 476,494-500], Auger processes[501-505] and multiplet (exchange) splittings[99,463,506]. We shall restrict here ourselves to a brief remark on the shake-up processes. Up to now we have treated simple photoionizations, i.e., ejections of a single electron, which may be characterized by the so-called Koopmans' configuration and which give rise to "main" bands in the electron spectra. A main peak may, however, be accompanied by satellite peaks of lower intensity that are due to more complex processes involving electron ionization and simultaneous electron transition from an occupied orbital to an unoccupied orbital. As a net result, one electron is ejected and the second is promoted to a virtual orbital. For example, the configuration $(1a_1)(2a_1)^2(1b_1)$ $(3a_1)^2(1b_2)^2(2b_1)$ for H_2O is a typical shake-up configuration for the 1s ionization. Actually, the term shake-up processes is used mostly in connection with ESCA, though the effect of simultaneous excitation need not be associated only with inner shell ionizations. The most straightforward theoretical approach to shake-up peaks is configuration interaction. Among the excited doublet states obtained, possible candidates for the shake-up states are those into which the Koopmans' configuration is mixed to an appreciable extent. The squares of the expansion coefficient standing at the Koopmans' configuration in the CI wave functions may be taken as relative intensities of the respective shake-up peaks. This is referred to as intensity borrowing from the main line. For valence orbitals one can usually identify the main line, which carries most of the spectral intensity. In the inner-valence shell region (approximately above 20 eV) it may be, however, difficult to make distinction between the main and satellite lines. With a series of molecules the shake-up processes were treated[498, 507-510] by the Green's function technique. As an example we present the case with the N_2 molecule[498]. For its $2\sigma_g$ ionization it was found that the main part of the intensity is distributed over four lines at 37.93, 39.94, 41.24 and 41.56 eV. In addition there is an intense line at 28.92 eV. It is therefore hardly possible to identify any among these lines as the "main" line corresponding to the $2\sigma_g$ orbital and the remaining as satellite lines. This sharing of the $2\sigma_g$ intensity is essentially a correlation effect and it means the breakdown of the one-to-one correspondence between the bands observed in the photoelectron spectrum and molecular orbitals.

The satellite lines may also be treated by means of MB-RSPT[238]. A very elegant formulation of the problem is possible by means of the quasidegenerate MB-RSPT[499] which permits all ionization potentials and their shake-up satellites to be treated simultaneously.

Finally, the basis set effect should be noted. We attempted to express the accumulated experience in a condensed form by means of Table 5.13.

5.H. Intermolecular Interactions

Knowledge of the behavior of atoms and molecules that participate in interactions is of primary importance for a further progress in many regions of physics, chemistry and biology. In this section we give an outline of basic methods of computation of interaction energies and their applications to some selected representative systems. Obviously, the choice of the optimum method and the basis set depends on the particular case, but some general guidance may still be formulated. A detailed information on the theoretical background of various approaches and their confrontation with experiment may be found both in classical[517,518] and recent[519,520] literature.

Since the pioneering work of London, it has been assumed for a long time that the perturbation treatment is the most suited approach to intermolecular interactions. Actually, use of perturbation methods is appropriate only with long range forces, where the electron exchange between the interacting systems A and B may be disregarded. Consider therefore first the case in which the distance between A and B is large. Then the wave function for a system AB may be written as

$$\Psi_0^{AB} = \Psi_0^A \Psi_0^B \qquad (5.27)$$

where Ψ_0^A and Ψ_0^B are wave functions for systems A and B, respectively.

T a b l e 5.13

Basis set effect on the calculated ionization potentials

Basis set	Approach	Notes on the use and accuracy achieved	Representative references
Minimum	any level of sophistication	not recommended	482, 511
DZ, DZ+P, extended	Koopmans' theorem	generally within 0.2–0.4 eV with respect to Hartree-Fock data; occasionaly over 1 eV for the DZ basis set	468, 482, 512, 513
DZ	ΔSCF, inner--shell IP's	1–2 eV within Hartree--Fock limit	514, 515
[5s3p/2s]	ΔSCF, inner--shell IP's	0.2–0.4 eV within Hartree-Fock limit	514, 515
[5s3p1d/2s1p]	ΔSCF, inner--shell IP's	~0.1 eV within Hartree-Fock limit	514, 515
DZ	reorganization and correlation effects accounted for	in average 0.5 eV within experiment	469, 482, 513
DZ+P	reorganization and correlation effects accounted for	~0.25 eV within experiment	482, 513
Extended	reorganization and correlation effects accounted for	compared to experiment maximum error 0.1–0.25 eV, depending on the basis set size and the particular molecule	469, 482, 513, 516

The total Hamiltonian for the system AB becomes

$$H^{AB} = H^A + H^B + V^{AB} \tag{5.28}$$

where H^A and H^B are Hamiltonians for systems A and B in the form of eqn. (1.1) and the interaction Hamiltonian, V^{AB}, is given by

$$V^{AB} = -\sum_{i\in A}\sum_{\beta\in B}\frac{Z_\beta}{r_{i\beta}} - \sum_{\alpha\in A}\sum_{j\in B}\frac{Z_\alpha}{r_{j\alpha}} + \sum_{i\in A}\sum_{j\in B}\frac{1}{r_{ij}} + \sum_{\alpha\in A}\sum_{\beta\in B}\frac{Z_\alpha Z_\beta}{R_{\alpha\beta}}$$

$$\tag{5.29}$$

Here indices i,j refer to electrons, indices α, β to nuclei and Z_α, Z_β are nuclear charges. With the wave function Ψ_o^{AB} and the Hamiltonian (5.28), the second-order perturbation theory gives the following expressions[519]

$$E^{(1)} = \langle \Psi_o^{AB} | V^{AB} | \Psi_o^{AB} \rangle \tag{5.30}$$

$$E^{(2)} = \sum_k \frac{|\langle \Psi_o^{AB} | V^{AB} | \Psi_k^{AB} \rangle|^2}{H_{oo}^{AB} - H_{kk}^{AB}} \tag{5.31}$$

The index k in eqn. (5.31) labels singly and doubly excited configurations within the frame of systems A and B (vide infra for further specification), and $(H_{oo}^{AB} - H_{kk}^{AB})$ are the respective excitation energies. $E^{(1)}$ represents the Coulomb electrostatic interaction energy (hereafter denoted as E_{Coul}) between the nuclear and electronic charge distributions given by wave functions Ψ_o^A and Ψ_o^B. In the orbital form it reads as follows[521]

$$E_{Coul} = -2 \sum_{a \in A} \sum_{\beta \in B} Z_\beta \left(a(1) \left| \frac{1}{r_{1\beta}} \right| a(1) \right) - 2 \sum_{\alpha \in A} \sum_{b \in B} Z_\alpha \left(b(1) \left| \frac{1}{r_{1\alpha}} \right| b(1) \right) +$$

$$4 \sum_{a \in A} \sum_{b \in B} \left(a(1)a(1) \left| \frac{1}{r_{12}} \right| b(2)b(2) \right) + \sum_{\alpha \in A} \sum_{\beta \in B} \frac{Z_\alpha Z_\beta}{R_{\alpha\beta}} \tag{5.32}$$

Here a and b are occupied MO's of systems A and B. Equation (5.32) is easily expressible in terms of integrals over atomic basis functions and elements of the density matrix. In eqn. (5.31) two terms may be distinguished. The first one is due to single electron excitations of the type $(a' \rightarrow r'')$ and $(b' \rightarrow s'')$, where a' and r'', respectively, are occupied and virtual MO's in the system A, and b' and s'' are occupied and virtual MO's in the system B. Contribution of these terms corresponds to the classical polarization interaction energy, E_p. Two-electron excitations $(a' \rightarrow r'', b' \rightarrow s'')$, i.e. simultaneous single excitations of either subsystem, may be taken as contributions to the second term – the classical London dispersion energy, E_D. If the Møller-Plesset partitioning of the Hamiltonian is used, E_D may be expressed in its simplest (orbital) form

$$E_D = 4 \sum_{a', r'' \in A} \sum_{b', s'' \in B} \frac{\left(a'(1) r''(1) \left| \frac{1}{r_{12}} \right| b'(2) s''(2) \right)^2}{\varepsilon_{a'} + \varepsilon_{b'} - \varepsilon_{r''} - \varepsilon_{s''}} \qquad (5.33)$$

where ε_k are the respective orbital energies. As shown by Kochanski[522] somewhat better results are obtained if the Epstein-Nesbet partitioning is used. In that case the denominator in (5.33) contains besides the orbital energies also the Coulomb and exchange integrals over orbitals a', r'', b' and s''. Computationally, evaluation of the expression (5.33) entailes only a small additional cost over the SCF calculations. It should be noted that E_D is inherently negative, because the nominator in (5.33) is positive and the denominator is negative. Expression (5.33) may be taken as an approximation to the intersystem correlation energy through second order. Obviously, it disregards the correlation energy change of each subsystem owing to interaction, i.e. the change in the intrasystem correlation energy, which should be taken into account at shorter distances. When dealing with long-range interactions, it is fair to note the multipole expansion of the interaction energy, i.e. the asymptotic expansion in power series of $1/r$. Convergence properties of this expansion were thoroughly examined by Ahlrichs[523]. The multipole expansion permits expressing the Coulomb, induction and dispersion interaction energy by means of the observables of individual subsystems such as charges, dipole, quadrupole and higher moments, molecular polarizabilities, excitation energies and ionization potentials. Explicit formulas are available in the literature, so it is not necessary to repeat them here. Only some specific expressions for the Coulomb interaction energy will be noted below.

Consider now the interaction energies at medium and short distances. Here the application of a standard perturbation theory is not free of complications. The main reason for this is the fact that the unperturbed wave function, $\Psi_0^{AB} = \Psi_0^A \Psi_0^B$, is antisymmetric only with respect to electron permutations in any subsystem but not to permutations in the whole supersystem AB. As the intersystem distance becomes shorter, the subsystem charge distributions begin to overlap and the intersystem electron permutations have to be taken into account. This may be easily done by assuming the antisymmetrized function $\alpha \Psi_0^{AB}$, but this function is not an eigen function of the Hamiltonian $H_A + H_B$ and it is

therefore not suited for the perturbation treatment. More recent methods, that are applicable to perturbation treatments at medium and
short distances, are discussed for example in Refs. 524-530. A discussion on this problem[531] with the connection to the asymptotic 1/r expansion[532] was reported by Kutzelnigg and Maeder. We restrict ourselves
here to a brief note that the introduction of the antisymmetrizer into
the wave function in the expression (5.30) gives rise to a first-order
exchange contribution, E_{ex}, which is always repulsive. On applying the
antisymmetrizer to the wave function in the second-order expression
(5.31) and taking into account the intermolecular excitations (i.e.
virtual MO's r" and s" in a particular molecule need not correspond to
its occupied orbitals), two further terms are obtained in addition to
the long-range terms - the charge-transfer term (from single electron
excitations) and the second-order exchange contribution (from two-electron excitations). It is evident that rigorous calculations at small
intersystem separations should be performed through higher than the
second order of the perturbation theory, but this is hardly feasible.

We comment now on the most widely used method of calculation of
intermolecular interaction energies which is based on the supermolecule
approach. In this approach it is necessary to calculate the energy of
the supersystem, E^{AB}, and the energies of the constituting subsystems,
E^A and E^B. The interaction energy is simply

$$\Delta E = E^{AB} - (E^A + E^B) \tag{5.34}$$

We may inquire which components of the perturbation expansion are included in ΔE. This depends on the particular method used for the calculation of E^{AB}, E^A and E^B. At the SCF level the interaction energy
ΔE_{SCF} is lacking the contribution of electron correlation, which is,
as we already know, equivalent to dispersion energy at large distances.
But otherwise ΔE_{SCF} gives a rigorous coverage of all other components
through all orders of the perturbation expansion. Evaluation of individual contributions requires, however, a special a posteriori decomposition of ΔE_{SCF}, in contrast to the perturbation treatment in which
all components are obtained directly from explicit expressions. The
decomposition[521,533] of ΔE_{SCF} may be performed in three steps. In the
first step the sum of the electrostatic and first-order exchange repulsion energies is obtained. This term, denoted usually as ΔE^1 =
$E_{Coul} + E_{ex}$, results from the antisymmetrized product of SCF wave func-

tions of the two subsystems, $\Psi_0^{AB} = a\Psi_0^A\Psi_0^B$. Computationally, ΔE^1 is obtained easily from the energy in the first iteration of the SCF calculation, if orthogonalized occupied SCF orbitals of isolated subsystems are used as starting vectors and if the energies E_{SCF}^A and E_{SCF}^B are subtracted. In the second step E_{Coul} may be obtained by means of eqn. (5.32). The difference $\Delta E^1 - E_{Coul}$ gives us E_{ex}. In the third step the SCF calculation is allowed to reach self-consistency and from the converged energy we obtain the second order contribution, $\Delta E^2 = \Delta E_{SCF}^{AB} - \Delta E^1$. The energy ΔE^2 is also commonly called the delocalization energy, E_{del}. It contains two main components: the polarization energy E_{pl} and the charge-transfer energy, E_{CT}'. The former may be viewed as the interaction which causes the mixing between the occupied and vacant MO's within each molecule, whereas the latter may be regarded as the interaction which causes intermolecular delocalization by mixing the occupied MO's of one molecule with the vacant MO's of the other and vice versa. Since the E_{pl} contribution may be evaluated directly[534], E_{CT}' is obtained by subtracting the term $(E_{Coul} + E_{ex} + E_{pl})$ from the total interaction energy. Nevertheless, E_{CT}' obtained in this way is not a genuine charge-transfer interaction energy E_{CT}, because it also contains further contributions, viz. the exchange polarization term, E_{expl}, and coupling interactions between components, E_{mix}. The most accurate scheme on the separation of all components of the interaction energy was developed by Morokuma and coworkers for both ground[534,535] and excited[536] states.

Physically meaningful is of course only the total interaction energy. Its decomposition to components is profitable, however, for the understanding of the nature of intermolecular interactions and also it is useful for the derivation of rules for the selection of a basis set. As an illustration of the complete decomposition of the total interaction energy, we present in Table 5.14 the data for the linear structure of the water dimer. The entries show a high sensitivity of the interaction energy components to the basis set used. In spite of this fact some general conclusions may be drawn about the relative importance of the energy components. Among the attraction contributions the most important one is E_{Coul}, followed by E_{CT}'. The total SCF interaction energy results from a complex interplay of all contributions from which no one can be disregarded. An interaction energy decomposition like that presented in Table 5.14 may be helpful in rationalizing the nature of bonding in hydrogen bonds. A detailed discussion on this topic may be found in the cited papers[538,540-542].

The entries in Table 5.14 permit us also to discuss some other

T a b l e 5.14

Energy decomposition for $(H_2O)_2$ with various Gaussian basis sets[a]

Basis set	Dipole moment $[10^{-30}$ Cm]	R_{OO} $[10^{-10}$ m]	ΔE_{SCF}	E_{Coul}	E_{ex}	E_{pl}	E_{CT}'	Ref
STO-4G	6.42	2.78	-26.4	-32.7	40.6	-1.1	-33.2	537
4-31G	8.67	2.98	-32.2	-37.3	17.5	-2.1	-10.3[b]	538
6-31G**	7.3	2.98	-23.4	-31.4	18.0	-2.1	-7.9	538
[541/31]	7.36	3.00	-19.8	-30.5	18.7	-3.1	-4.9	537
HF limit	6.65[c]		-15.4					539

[a] If not otherwise noted, the entries are in kJ/mol; the dipole moment refers to the monomer.
[b] In a more detailed decomposition[535] this energy may be further split to E_{CT} = -8.8, E_{expl} = -1.7, and E_{mix} = 0.2 kJ/mol.
[c] This value (see Table 2.15) and ΔE_{SCF} were obtained with different basis sets.

problems. Consider first the total interaction energy. It may be noticed that with small basis sets the total interaction energy is overestimated. We mean the overestimation with respect to larger basis sets. A reliable experimental value is not available (for a comparison of theory with experiment see Ref. 169). The overestimation is mostly due to the so called counterpoise or superposition error[543-547], which originates from the fact that if the subsystem A (B) is represented by a small basis set, it tends to improve the quality of its basis set by means of the basis functions located on the other interacting system B (A). This brings about an energy lowering in the system AB, which results in the overestimation of the interaction energy, in particular of its E_{del} component. The magnitude of this effect may be easily estimated. It is sufficient to compute the SCF energy of the system A, E_c^A, with the basis set and at the geometry used for the supersystem AB but with the zero charges on nuclei that constitute the molecule B. The energy E_c^B for the system B is then obtained analogously. The calculations of this type are not associated with any considerable additional cost, because the list of two-electron integrals is the same as with the supersystem. Actually only the nuclear attraction integrals have to be recalculated because of zero charges on nuclei bearing the "ghost" orbitals. The interaction energy thus becomes $\Delta E = E_{SCF}^{AB} - (E_c^A + E_c^B)$ instead of $\Delta E = E_{SCF}^{AB} - (E_{SCF}^A + E_{SCF}^B)$. The difference $\Delta \varepsilon = (E_{SCF}^A + E_{SCF}^B) - (E_c^A + E_c^B)$ represents the counterpoise correction. As the size of the basis set is extended, its importance decreases. Extremely high counterpoise correction of 24 kJ/mol is found[545] for the water dimer with the STO-3G basis set. In this case $\Delta\varepsilon$ is higher than the dimerization energy itself. Evidently, the STO-3G basis set is not suited for the application to $(H_2O)_2$, though the counterpoise correction computed in this way should merely be taken as an upper limit to the effect of basis set extension[545]. With the optimized minimum contracted Gaussian basis set $\Delta\varepsilon$ reduces[548] to 4.7 kJ/mol. With the 4-31G basis set it becomes[549] about 4 kJ/mol and with the [541/31] basis set it is of about[169] 3 kJ/mol. A deeper insight into this problem is achievable by means of the perturbation approach[528] which also permits treatment of the basis set effect on the dispersion energy[550,551]. Counterpoise error was also examined with CI calculations[169,552]. Actually, the basis set should be tested with respect to the counterpoise correction in any type of calculation of intermolecular interaction energies.

There is also another reason for the overestimated interaction energy in $(H_2O)_2$. With small and medium sized basis sets the dipole moment of the water molecule is overestimated, which brings about a too high value of E_{Coul}. This may be demonstrated by means of the multipole expansion truncated to its first term

$$E_{es}^{mult} \approx \frac{\mu^2}{R^3} \tag{5.35}$$

where R is the intermolecular distance and μ is the dipole moment. This view is however rather oversimplified. As suggested by the data in Table 5.14, the relationship between E_{Coul} and μ is more complex. Specifically, two closely related effects should be considered: (a) the inclusion of further terms into the multipole expansion, and (b) penetration of charge distribution of the two systems at short distances. For a discussion of these effects we make use of results of our calculations[553] on a somewhat uncommon system $(BH)_2$. For the BH molecule, the DZ basis set gives a remarkably high quadrupole moment $\theta_{zz} = -13.467 \times 10^{-40}$ Cm^2 (computed with respect to the center of mass). The dipole moment and the octupole moment are predicted with the same basis to be $\mu = 6.266 \times 10^{-3}$ Cm and $O = -0.439 \times 10^{-50}$ Cm^3. The dipole moment is in fair agreement with the value of 5.780×10^{-30} Cm given by the extended basis set[554]. We now substitute the DZ data into the multipole expansion of the electrostatic interaction energy in the point multipole approximation. We assume two orientations. For the linear structure the electrostatic interaction energy may be expressed as

$$E_{es}^{mult} = -\frac{2\mu^2}{R^3} + 6\frac{Q^2}{R^5} - 8\frac{\mu O}{R^5} = E_{es}^{\mu\mu} + E_{es}^{QQ} + E_{es}^{\mu O} \tag{5.36}$$

whereas for the perpendicular structure it becomes

$$E_{es}^{mult} = \frac{3}{2}\frac{\mu Q}{R^4} - 3\frac{Q^2}{R^5} = E_{es}^{\mu Q} + E_{es}^{QQ} \tag{5.37}$$

In Figs. 5.6a and 5.7a they are compared with E_{Coul} obtained by means

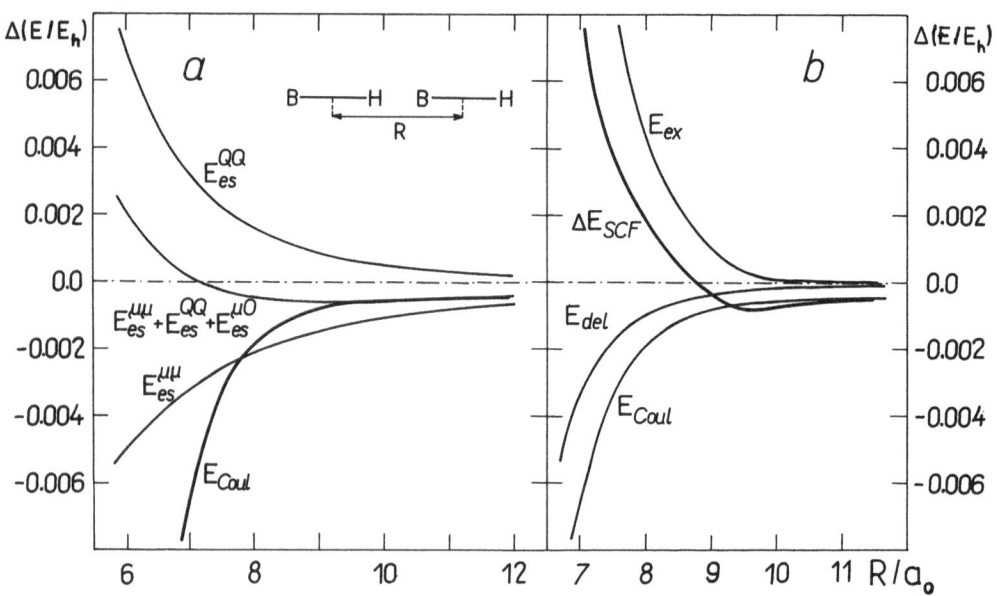

Figure 5.6

Interaction BH-BH in the linear configuration[553]. a) Electrostatic interaction energy and the first terms of its multipole expansion. Intermolecular distance, R, and quadrupole and octupole moments refer to the centers of mass. b) SCF interaction energy and its components.

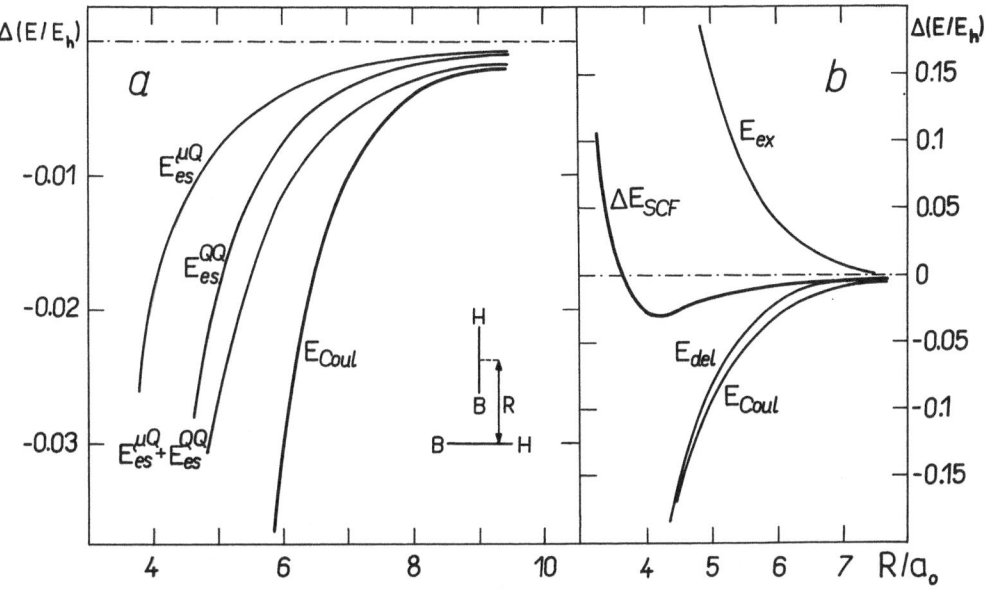

Figure 5.7

Interaction BH-BH in the perpendicular configuration[553]. a) Electrostatic interaction energy and the first terms of its multipole expansion. Intermolecular distance, R, and the quadrupole moment refer to the midpoints of bonds. In this case the quadrupole moment is -19.891×10^{-40} Cm^2. This choice of the origin of coordinates does not affect qualitatively any conclusion noted in the text. A detailed discussion on the problem of selecting the coordinate center was given by Lischka[315]. b) SCF interaction energy and its components.

of eqn. (5.32) and Figs. 5.6b and 5.7b present the total interaction energies ΔE_{SCF} and their components. A deeper minimum on the ΔE_{SCF} curves was found for the perpendicular configuration. It is located at R = 4.2 and its depth is $\Delta E_{SCF} = -0.029198$ E_h. With the linear configuration the minimum is very shallow ($\Delta E_{SCF} = -0.00058$ E_h) and it is located at the distance as long as $R \approx 10$ a_o. Just the opposite order of stabilities might be expected, if the prediction were based solely on the R^{-3} term. This term, which may be assumed to be the dominant component among the long-range forces in the interaction of two dipoles, is vanishing with the perpendicular configuration, so that the linear structure should be favored. The origin of this incorrect prediction may be understood from Fig. 5.6a. Since the quadrupole moment of BH is high, the repulsive term E_{es}^{QQ} compensates the term $E_{es}^{\mu\mu}$, or even dominates over $E_{es}^{\mu\mu}$ at distances $R < 7$ a_o. $E_{es}^{\mu O}$ is considerably smaller than $E_{es}^{\mu\mu}$ and E_{es}^{QQ} at any distance. The total electrostatic interaction energy in the linear structure is therefore very small and the bonding in this system resembles the van der Waals interaction. With the perpendicular configuration, the two most important terms, $E_{es}^{\mu Q}$ and E_{es}^{QQ}, are also comparable in magnitude but they are of the same sign. Figs. 5.6a and 5.7a may be taken as an example showing how the knowledge of first terms of the multipole expansion may be useful in establishing the dependence of the interaction energy on the orientation of molecules. The fact that the treatment cannot be restricted to the very first term of the expansion has been recognized long time ago [555]. Another feature of Figs. 5.6 and 5.7, which should be commented on, is a considerable difference between the E_{Coul} and E_{es}^{mult} curves with both $(BH)_2$ configurations. It should be realized that the multipole expansion is in fact a long-range approximation and that it diverges at short distances. The convergence is possible only if the charge distributions of both subsystems do not penetrate into each other. Owing to penetration[521,556], E_{Coul} is more attractive than E_{es}^{mult}; the two energies approach asymptotically only at large distances. From Fig. 5.6a we can see that even if the first two terms in the multipole expansion indicate the convergence (i.e. $E_{es}^{\mu\mu} > E_{es}^{QQ} + E_{es}^{\mu O}$ for linear configuration), E_{es}^{mult} and E_{Coul} still differ considerably. Besides the penetration effect, this also may be assigned to an inevitable truncation of higher terms in the multipole expansion. It should be noted at this point that the convergence of the multipole series may be improved. Various approaches based on the decomposition of the molecular charge density into smaller distributions and procedure to generate high-order moments were suggested by Mezei and Campbell[557].

A similar analysis of the long-range behavior may also be performed for other components of the interaction energy. With the polarization energy, it is necessary to know besides the multipole moments also the polarizabilities of molecules, whereas an analysis of the dispersion energy requires an extra knowledge of polarizabilities, excitation energies, or ionization potentials, depending on the approximation used. Many among these quantities are not accessible experimentally or their experimental values are open to large uncertainties, so that accurate ab initio calculations become very topical here. As an example of a thorough analysis of the interaction energy we cite a few selected papers: Schuster and coworkers[558] treated the systems Li^+-H_2O and Li^+-H_2CO; Staemmler[559] the Li^+N_2 system; Kutzelnigg and coworkers[312] the system Li-H$_2^+$; Lischka reported the calculations for He-HF and He-H$_2$O[314,315] and for H$_2$-HF and H$_2$-H$_2$O[315]; and Garrison and co-workers[460] for the He-H$_2$CO system. A detailed analysis of the van der Waals complexing of two ethylene molecules was reported by Wormer and coworkers[561,562].

The methods discussed above in this section may also be applied to very large systems, though the calculations are feasible only with a minimum basis set. A typical example of the theoretical approach of this kind is the study of interactions of cations with biomolecules as it is made in the laboratory of A. and B. Pullman[563,564]. The system biomolecule-cation is treated as a supermolecule and the calculation is repeated for many different distances and orientations of the cation with respect to the biomolecule.This permits to find preferred positions of the cation binding. Fig. 5.8 presents preferred positions

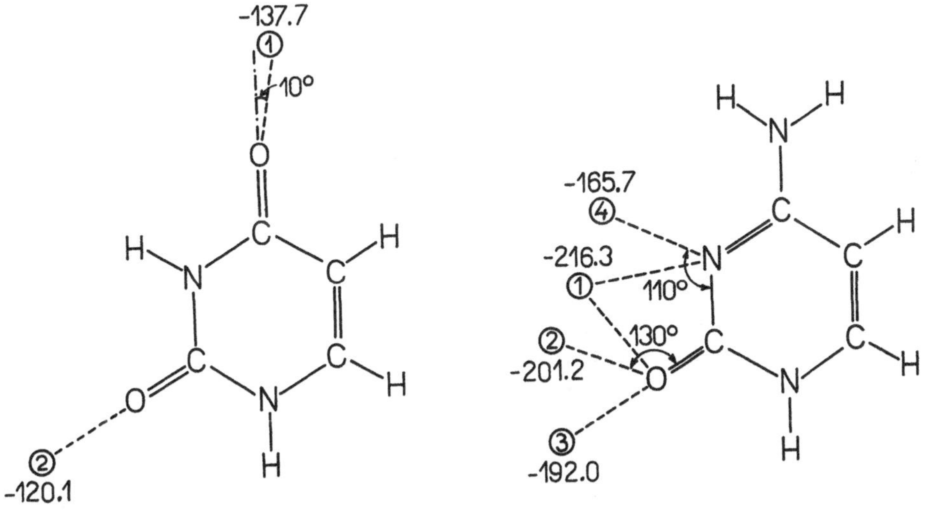

Figure 5.8

SCF interaction energies (in kJ/mol) of Na$^+$ with uracil and cytosine given by a minimum basis set[564]. The preferred locations of Na$^+$ are denoted by circles with numbers that give the order of their preference.

for the Na$^+$ binding with uracil and cytosine. The results of this type on the optimum cation locations and the order of the respective interaction energies permit to draw conclusions about cation binding in

real biological systems such as the nucleic acids. The decomposition
of the interaction energy of Na^+ with uracil and cytosine showed that
E_{Coul} is the dominating term and that the other two components, E_{ex}
and E_{del}, tend to cancel each other at the optimum distance of Na^+
from the binding site. In such a situation where the cation binding
is governed by E_{Coul}, it is profitable to make use of the molecular
<u>electrostatic potential map</u>. This approach was introduced by a group
of Italian workers[565]. Its essence lies in the calculations of potentials $V(\vec{r}_i)$ generated by the molecule at the points i. The interaction
energy with a point charge, q_i, located at the point i is approximated
simply by a product $q_i(\vec{r}_i)V(\vec{r}_i)$. The cost associated with the construction of molecular electrostatic potential maps is relatively low
because the computation of the whole supersystem is avoided. One cannot of course make unequivocal predictions about the preferred binding
sites for cations on the basis of electrostatic potential maps only.
This technique should be rather taken[564] as a tool giving us information about the positions that should be selected for a complete SCF
supermolecule calculation.

The versatile utility of the electrostatic potential maps for qualitative predictions may be demonstrated by an application of another
kind that was reported by Kollman[566]. He was able, on the basis of the
quadrupole-quadrupole point charge model, to give the explanation for
why the most stable structures of Cl_2-Cl_2 and F_2-F_2 dimers are
"L-shaped", in contrast to H_2, N_2 and other homonuclear diatomics that
are T-shaped. From Fig. 5.9 it is seen that the most positive locations of the potential for H_2 are along the bond axis, whereas the
most negative areas are perpendicular to this molecule, bisecting the
H-H bond. This suggests that the T-shaped structure would be predicted
to bring the most positive end of one molecule toward the most positive end of the other. With Cl_2 the most positive potential is along
the bond axis. The most negative potential is almost straight above a
Cl atom, but inclined at $\sim 15°$ relative to the axis. This direction
corresponds almost exactly to the angle between Cl_2 molecules in the
optimum structure of $(Cl_2)_2$ given by SCF calculations.

As with other applications treated in Chapter 5, the rest of this
section will be devoted to the problem of correlation energy and to
the basis set effect. Let us first try to find types of interactions
where the correlation effects are almost negligible, where the estimation of the correlation energy is profitable, and where its calculation is unavoidable. For this purpose we set up Table 5.15. It col-
lects the results for a variety of systems, obtained mostly with large

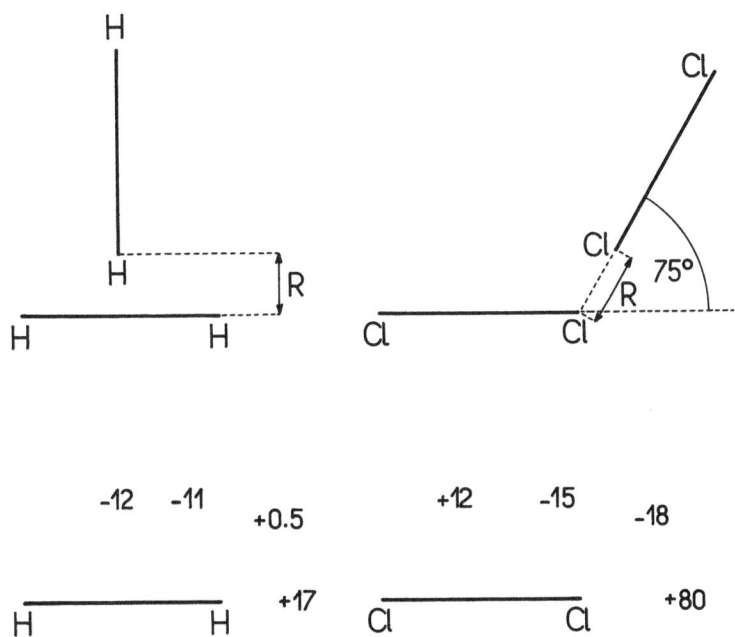

Figure 5.9

Optimum geometries[566] (at the top) for H_2 and Cl_2
and the electrostatic potentials for $R = 5a_0$ (at
the bottom).

basis sets. For the simplicity we shall ignore the fact that the correlation energies were calculated by different methods. An immediate
observation from Table 5.15 is that in interactions of univalent ions
with a polar molecule (Li^+-H_2O, F^--H_2O), the correlation contribution
represents only a very small fraction of the total interaction energy.
Correlation effects also appear to be small in interactions of the Li^+
cation with a nonpolar molecule. Here, however, the correlation contributions depend strongly on the configuration assumed. As documented
by the results for Li^+-N_2, they may even differ in sign for two different configurations. An unusual type of the interaction is provided
by the example of the system lithium atom - water molecule. Also here
the effect of the correlation energy is relatively small, but the high
SCF interaction energy is somewhat surprising. A remarkable feature of
these calculations is the fact that they predict[567,570,571] a novel
class of molecular complexes of which Li-H_2O may be considered the pro-

T a b l e 5.15

SCF and correlation contributions to interaction energies[a] (in kJ/mol)

System	ΔE_{total}	ΔE_{SCF}	ΔE_{corr}	Ref.
Li^+-H_2O	-146.11	-150.96	+4.85	169
F^--H_2O	-109.46	-101.34	-8.12	169
Li^+-N_2[b]	-62.11	-56.68	-5.43	559
Li^+-N_2[c]	-6.09	-12.75	+6.66	559
Li^+-H_2	-24.94	-23.97	-0.97	312
$Li-H_2O$	-56.7	-51.0	-5.7	570
H_2O-H_2O	-25.31	-21.51	-3.80	169
HF-HF	-14.08	-14.47	+0.39	316
He-HF	-0.630	-0.171	-0.459	314
$He-H_2O$	-0.614	-0.234	-0.380	314
H_2-HF	-2.568	-1.494	-1.074	315
H_2-H_2O	-4.372	-2.429	-1.943	315
Ne-Ne	-0.291	0.118	-0.409	568
H_2-H_2	-0.079	0.231	-0.310	569

[a] The calculations were performed at near-optimum geometries, for details see cited papers; ΔE_{corr} are inferred from the published ΔE_{total} and ΔE_{SCF} data.
[b] Linear configuration.
[c] Perpendicular configuration.

totype. The papers quoted demonstrate the predictive power of ab
initio calculations which, in this case, provide a challenge to cros-
sed-beam experimentalists.

The correlation energy effect may be of some importance with weak
hydrogen bonds. While it is unimportant with the system HF-HF (vide
infra), it constitutes about 15% of the total interaction energy with
H_2O-H_2O. In spite of this fact, the SCF calculations provided many interesting results on various hydrogen bonded systems, in both ground
538,540-542 and excited states (see for example Refs. 536,542,572).
To some extent, the agreement of SCF interaction energies and some related quantities with experiment is due to cancellation of several effects: neglect of the correlation energy, disregarding the zero-point
energy, or the basis set effect. A more rigorous comparison with experiment will be presented in Section 5.I. for some selected systems.

With van der Waals complexes, such as for example H_2-H_2, the SCF approach is inherently incapable of giving a minimum on the potential curve[518] and the inclusion of the correlation energy becomes inevitable. At the first step, it appears natural to include only the intersystem correlation energy which, as we already know, becomes asymptotically equivalent to the dispersion energy (eqn. (5.33)) at large distances. This means that the total interaction energy might be approximated by the sum $\Delta E_{SCF} + E_D$. This approach was tested recently by Prissette and Kochanski[568] with the system Ne-Ne. The obtained depth of the van der Waals minimum represented about 71% of the value derived from experimental data. A direct comparison of the dispersion energy with ΔE_{corr} (obtained from CI-SD calculations) was performed[569] with the H_2-H_2 system. The results showed a qualitative agreement between the two approaches. Since the calculation of the E_D term is relatively simple, it is also feasible with somewhat larger systems. For example, Umeyama and Morokuma[573] corrected in this way SCF interaction energies of halogen complexes, though the basis set used was lacking polarization functions so that only a small fraction of the dispersion energy was recovered. As shown by Kochanski[522], it is necessary in such an approach to make use of a basis set containing polarization functions with a very low exponent. Such functions are, however, not suitable for calculation of the SCF interaction energy, since they give rise to a large counterpoise error[546]. This problem is soluble in two ways. Either the counterpoise correction is taken into account, or two sets of polarization functions are used. In the latter case, one set of polarization functions ensures correct SCF energies whereas the other is present to bias E_D. For example, the suitable exponents[574] for the hydrogen GTF's are 1.0 and 0.1.

From an examination of Table 5.15 it may be noticed that three values of ΔE_{corr} have a positive sign. All these three values were obtained by CI calculations (Li^+-H_2O) or by CEPA (Li^+-N_2, HF-HF), i.e. by methods that besides the attractive intersystem energy also account for the change in the intrasystem correlation energy. Since the latter may be positive or negative, it may in some instances compensate or even overweigh the intersystem contribution. Needless to say that in such a case the dispersion energy is meaningless. Origin of the positive sign with ΔE_{corr} may be understood from the correlation effect on the properties of subsystems. For example, with the HF molecule[316,542] correlation energy reduces the dipole moment and increases the polarizability, which brings about a lower Coulomb energy and higher polarization and dispersion energies. The balance of all these changes determines the magnitude and sign of the total ΔE_{corr}. Furthermore, the interaction energy may be affected by the coupling of inter- and intrasystem correlation effects. Since the dominant part of the change in intrasystem electron correlation is accountable for by double excitations in the frame of subsystems and the dispersion energy (the dominant part of the intersystem correlation energy) by simultaneous single excitations in each subsystem, the inter-intra correlation coupling is amenable to examination only if triply excited states are included. The problems of the inter-intra correlation coupling were discussed by Kutzelnigg[529]. Because of complexity of calculations the numerical results are, however, still scarce (e.g., Refs. 552a,575, 576). It should be noted that calculations of this type are meaningful only if a large saturated basis set is used. Otherwise the coupling effect might be obscured by the counterpoise error.

T a b l e 5.16
Basis set effect on the calculated interaction energies

Type of inter-action	Basis set	Notes on the use and the accuracy achieved	Representative references
Ion-dipole or hydrogen bond	minimum or DZ	Structures correctly estimated, distances underestimated (by ~10%). Interaction energies overestimated (120-150% or more of experimental energies. Accuracy depends mainly on the estimated multiple moments and polarizabilities of particular molecules and on the counterpoise correction for a given basis set)	89, 578-580
	DZ+P	Mostly overestimation of interaction energies (120-130% of experimental values)	537, 539, 578
	extended	Essentially agreement with experiment	see for them Section 5.I.
van der Waals complexes	extended	Only large basis sets with polarization functions are to be recommended; inclusion of correlation effects unavoidable	216, 522, 546

Finally, the basis set effect should be commented on. This is, however, a rather troublesome task because comparison of results given by different basis sets with experiment is very difficult. This is due to the lack of systematic theoretical studies, but also due to low reliability of many experimental data. The comparison is particularly

difficult if the interaction is dependent on the orientation of subsystems. Next, the compatibility of theory with experiment is complicated by zero-point energies and entropy effects[577]. Nevertheless we
attempted in Table 5.16 to outline some trends.

5.I. Solvation

Up to now our attention was mainly devoted to calculations of the
energy and other quantities referring to free, isolated molecules. The
computational techniques and their applications were demonstrated to
be profitable in the exploration of physico-chemical properties of
free molecules and their reactivity in the gas phase (thermodynamic
functions, equilibrium and rate constants). However, the gas-phase
processes represent only a special minor part of chemistry. Not only
processes in biological systems, but also processes in laboratory conditions proceed typically in the liquid phase - or expressed more specifically - in the solution. It is therefore not surprising that the
effort for applications of ab initio calculations is also still increasing in this very important field[519,581-584].

From the theoretical point of view, the system we are going to in-
vestigate is constituted by a large number of solute and solvent molecules. The problem to be treated is therefore inherently of the many-
-body nature and its analytical solution is inaccessible. Inevitably,
we are forced to adopt some simplifying view. Since we are only dealing
with ab initio calculations, we find it convenient to make use of the
following classification of the existing theoretical approaches:

(i) Calculations of pair interaction energies and their use in a
statistical thermodynamic treatment.

(ii) The "supermolecule" approach.

(iii) Methods accounting for the solvent effect by means of the continuum model.

(iv) Methods combining the continuum and supermolecule approaches.

First we make an attempt to break through the solvation phenomena
by means of the pair interactions. The underlying theory for the two-
-body interactions was dealt with in the preceding section, so that
we may immediately pass to the problem of testing the predicted interaction energies. Recent progress in experimental techniques resulted
in accumulation of thermochemical data, against which the predicted interaction energies may be judged. Especially, the data on the solvation of monoatomic ions with a single water molecule[585] are well suited
for this purpose. Their use is shown in Table 5.17, where they are

T a b l e 5.17

Theoretical and experimental heats of hydration of selected ions (in kJ/mol)

Complex	ΔE_{SCF}	ΔE_{corr}	Zero-point correction	$\Delta H^o_{298} - \Delta H^o_0$ (calcd.[a])	ΔH^o_{298} Calcd.	ΔH^o_{298} Exptl.[585]
Li^+-H_2O[b]	-147.3	-1.1	8.5	-4.4	-144.3	-142
Li^+-H_2O[c]	-151.0	4.9	8.6		-137.5[d]	-142
Na^+-H_2O[b]	-100.2	-1.1	6.3	-3.5	-98.5	-100
K^+-H_2O[b]	-69.6	-1.4	5.5	-3.1	-68.6	-70.7
F^--H_2O[b]	-99.2	-3.1	13.3	-4.7	-93.7	-97.5
F^--H_2O[c]	-101.3	-8.1	13.3		-96.1[d]	-97.5
Cl^--H_2O[b]	-49.6	-2.4	6.6	-3.2	-48.6	-54.8

[a] Recalculated[586] from original[267] geometries and vibrational frequencies.
[b] Ref. 267.
[c] Ref. 169.
[d] ΔH^o_0 value.

compared with the results of selected ab initio calculations performed with very large CGTF basis sets. The overall agreement is excellent, though it is open to some uncertainties. It should be noted that the experimental ΔH^o_{298} are estimated[585] to be within 4-12 kJ/mol. With the Li^+-H_2O complex, ΔH^o_{298} was not obtained directly from experiment but merely by the extrapolation[587] of data for higher associates $Li^+-(H_2O)_n$ with n = 2-6. Also the theoretical treatments presented in Table 5.17 are not free of feeble points. The zero-point and $\Delta H^o_{298}-\Delta H^o_0$ corrections were calculated within the rigid rotor-harmonic oscillator approximation. With the complexes of the X^--H_2O type, however, the use of harmonic frequencies for the intersystem vibrational modes is questionable, because of considerable anharmonic behavior of the potential surface near the minimum. Uncertainty is also associated with the correlation energy contributions. Kistenmacher et al.[267] estimated them by means of the semiempirical approach based on Wigner's model. As noted in Section 4.C. such an approach can provide merely semiquantitative predictions. The ΔE_{corr} data obtained by Diercksen et al.[169] from CI-SD calculations are certainly more realistic. In contrast, the

SCF energies coming from the two laboratories are highly reliable and they are close in absolute value. The former[267] are practically at the Hartree-Fock level, the latter[169] were obtained with only a slightly smaller basis set. With smaller basis sets, the differences between the respective ΔE_{SCF} values are much larger and in such a case it is just ΔE_{SCF} which is the main source of error in the estimated ΔH. This may be taken as reasoning why in the treatments of that type the corrections to ΔE_{SCF} are mostly disregarded and ΔE_{SCF} is considered directly a semiquantitative estimate of ΔH_{298}^o. Since the correction terms tend to cancel out (see Table 5.17), this approach is not so crude as it might appear. As regards the effect of ΔE_{corr} alone, we found already in Table 5.15 that it is small with this type of complexes. Nevertheless it is by no means negligible when highly accurate predictions are attempted: with the F^--H_2O complex it represents about 8% of the SCF interaction energy, and in the interaction OH^--H_2O it is as high[588] as 15 kJ/mol and it represents about 15% of the SCF interaction energy.

Assume now that we are in the position to be able to calculate reliable pair potentials. The next task is to employ some statistical thermodynamic model which would permit us to pass from the pair complexes solvent-solute and solvent-solvent to a real liquid. A theoretical analysis of this problem is beyond the scope of this book, so that we restrict ourselves to stating that, in the conjuction with ab initio calculations, the most sophisticated approach appears to be the statistical mechanics computer simulation of the finite system of N molecules in the volume V at the temperature T. The essence of the calculation is geometry configurational averaging of the system by means of the Monte Carlo method[589,590]. It is necessary to know the configurational interaction energy of N particles

$$V(X^N) = \sum_{i<j}^{N} V_{ij}^{(2)}(X_i,X_j) + \sum_{i<j<k}^{N} V_{ijk}^{(3)}(X_i,X_j,X_k) + \dots \qquad (5.38)$$

for which usually the pair-wise additivity is assumed, which means that the expansion (5.38) is truncated at the $V^{(2)}$ terms. $V^{(k)}$ terms in eqn. (5.38) represent the k-body contributions to the interaction and X_i represents the configurational coordinates of the i-th particle. The probability of a given geometry configuration is proportional to the Boltzman weighting for that configuration. The average distri-

bution of molecules in a liquid is usually represented by radial distribution functions, from which one may obtain by integration the
number of particles inside a shell of radius r about a given particle.
Since the number of configurations generated in the Monte Carlo simulation is of the order 10^5-10^6, the configurational interaction energies must be easily expressible. This is achieved by calculating a
sufficiently large number of pair interaction energies at different
separation and orientation of the two molecules and by fitting the obtained energy surface by an analytical function of a suitable form.
The use of nonempirical potentials in this way[168] in Monte Carlo calculations was pioneered and systematically developed[584] by Clementi
and his group. Since the applications and results obtained were reviewed[584] recently, we note only briefly on some of them. A critical
test for this theoretical approach was provided by the treatments on
liquid water[266,539,591-594]. In the cited papers the calculated radial distribution functions were thoroughly compared with the experimental ones. The agreement was fair at the Hartree-Fock level, the
discrepancies being reduced considerably if the correlation energy
was accounted for. This suggests that the interaction potentials
given by ab initio calculations are realistic. The results obtained
with different potentials for internal energy and specific heat are
summarized in Table 5.18. To make the CI-SD data compatible with the

T a b l e 5.18
Internal energy and specific heat of water given by the Monte
Carlo calculations with 64 molecules for 298 K and with the ex-
perimental density of liquid water

Potential	U [kJ/mol]	C_v [J/deg mol]	Ref.
Hartree-Fock (HF)	-28.9	75.3	266
HF + London dispersion term	-38.5	62.8	266
HF + Wigner-type estimate of E_{corr}	-33.5		266
HF + dispersion term (perturbation calcn.)	-30.1	71.1	591
HF + CI-SD[592]	-28.5[a]	75.3[a]	593
Experiment	-33.9	74.9	

[a] 343 water molecules taken in the Monte Carlo calculation.

other entries of Table 5.18 we present the results[591] obtained with the Hartree-Fock potential corrected by the dispersion term (calculated perturbationally) for different numbers of treated water molecules. For 64, 125 and 343 molecules of water per unit cell, the computed internal energies were -30.1, -28.5 and -27.6 kJ/mol, respectively. The internal energy appeared to be more sensitive to the potential used than the predicted structure of liquid water. A rather large effect of the correlation energy on the internal energy is not surprising, inasmuch as the correlation energy contribution in the optimum configuration of the water dimer represents about 15% of the SCF interaction energy (see Table 5.15; a detailed comparison of ab initio calculations and experiment for the water dimer is reported in the cited papers[169,595]).

Remarkable Monte Carlo results were obtained for the lithium fluoride ion pair in water[596,597]. With highly accurate potentials water-water[539,595] and ion-water[168,598] including also three-body terms, distributions were obtained for the oxygen and hydrogen atoms of water molecules around the ion pair (for a fixed Li-F distance taken as a parameter). Two and three dimensional representations of these distributions gave fascinating pictures of the structure of water with a clear-cut first hydration shell and a loosely bound second hydration shell. Monte Carlo calculations also permitted to determine coordination numbers for Li^+, Na^+, K^+, F^- and Cl^- ions[584,599,600].

Encouraging results noted above prompted the Clementi's group to extend the activity to a very complex problem - determination of water's structural organization around biomolecules[584,601-606]. The crucial problem, again, is obtaining the pair interaction potentials. In this case these are the biomolecule-water and water-water potentials. As with simpler systems noted above, the pair potentials must be determined from a sufficiently large number of geometry configurations. For example, in the study of the interaction of water with 21 amino acids[601], altogether 1690 interaction energies were computed and in a similar study for four DNA bases[602] altogether 368 configurations were computed. The biomolecule-water interaction energies obtained were fitted with an analytical expression of the following form[601,607]

$$V^{(2)} = \sum_{i<j} v_{ij}^{ab} = \sum_{i<j} \left(-\frac{A_{ij}^{ab}}{(r_{ij})^6} + \frac{B_{ij}^{ab}}{(r_{ij})^{12}} + \frac{C_{ij}^{ab} q_i q_j}{r_{ij}} \right) \qquad (5.39)$$

which contains the Lennard-Jones potential and the Coulombic term. A, B and C are fitting constants, r_{ij} is the distance between an atom i on the solute molecule and j is an atom of the water molecule, and q are atomic net charges (given in this case by the Mulliken population analysis). Besides the computational simplicity, the selection of the particular form (5.39) also aimed at achieving, to some extent, trans- ferability of the analytical pair potentials to other molecules that are chemically similar to those previously studied. For this reason additional indices a and b were introduced to eqn. (5.39). These not only distinguish between atoms of different Z value (for example a hydrogen atom from a carbon atom) but also, within a group of atoms of equal atomic number Z, differentiate the "electronic environment" of an atom in the molecule. Transferability of potentials was tested[603] by applying the pair potentials, obtained for 21 amino acids[601], to phenylalanine, which was not included in the treated set. The approx- imate interaction energies obtained in this way were then compared with the results of direct calculations on the complex phenylalanine- -water in 75 different configurations. The agreement was satisfactory, mostly within ~4 kJ/mol. Once a library of parameters for different "classes" of atoms is available, the pair potentials of this type be- come very informative for the investigation of solvation of biological molecules. A circumstance that this approach is based on nonempirical calculations is very important, because for biomolecules in solution it is rather difficult to select and often even find a sufficient sample of experiments of direct relevance and equal reliability to be able to infer the empirical potential functions.

Supermolecule model. By a "supermolecule" we imply a model con- sisting of the solute molecule surrounded by a certain number of sol- vent molecules. Pair complexes solute-solvent and solvent-solvent may be considered the simplest supermolecules. Since the cost of the su- permolecule approach becomes prohibitive as the number of solvent mol- ecules is increased, in most treatments only the first solvation shell is assumed. Such small clusters cannot of course provide a realistic model of a liquid but rather they give us a theoretical picture of what is referred as to "the solvation in the gas phase". As with the approach dealt with in the last paragraph, the ab initio calculation on the supermolecule should be followed by a statistical thermodynamic treatment. The use of the standard statistical thermodynamic is straightforward, in which case the supermolecule approach becomes e- quivalent to treatment of common chemical equilibria dealt with in Sec- tion 5.F. The calculations presented in Table 5.17 are just of this type. It should be noted, however, that the supermolecule approach is justifiable only if the pair interactions are strong enough and for relatively low temperatures. Hence, if the structure of the second solvation shell or the structure of weakly solvating solvent is to be determined,for example,the Monte Carlo calculations should be preferred. Nevertheless it is fair also to note the virtues of the supermolecule approach. The latter permits a microscopic view into the interaction, particularly the partitioning of the interaction energy (see the preceding section), and an explicit examination of the effect of the solute molecule on solvent molecules and vice versa. Another point in favor of the supermolecular approach is that besides the energy, it also provides the wave function. As will be shown later on, this permits us to give an account of changes in many different properties of molecules upon solvation. We first comment, however, on the problem of many-body interactions. Since the supermolecule approach includes them explicitly, it permits estimation of the effect of their neglect in the approaches based on pair potentials.

It is convenient to discuss the three- and higher body interaction energies with the clusters of water molecules because for these systems a series of ab initio calculations was reported. The first of them, with the minimum STO-4G basis set, was reported by Del Bene and Pople[608] in 1970. Shortly after it calculations with extended basis sets followed[595,609,610]. The cited papers were not limited to the calculation of the interaction energy but they also concerned the problem of the optimum geometry of $(H_2O)_n$ polymers - in particular the problem whether ring or open structures are the optimum configurations. Nowadays, it appears to be established that with $(H_2O)_n$ clusters for $n \lesssim 5$ the ring structures are favored[595]. With $(H_2O)_3$, for example, the optimum structure is close to the C_3 symmetry with the O-O bond lengths round 2.96×10^{-10} m in an almost equilateral triangle. Search for optimum structures of $(H_2O)_n$ clusters with $n \gtrsim 3$ rapidly becomes troublesome as n is increased. Even though the geometries of water molecules are kept rigid, the number of intermolecular bond lengths and possible mutual orientations of molecules is too high. One compares therefore mostly only the stabilities of structures selected by intuition which, of course, may lead to erroneous predictions. With large clusters one finds typically a large number of structures that are close in energy. Some of them need not be real minima or not even saddle points on the respective energy surface. Furthermore, the order of stabilities may be considerably affected by zero-point energies. Finally it must be realized that the stability is governed by free enthalpy

and not by the energy. The last circumstance is very important[577,586];
with the water dimer, for example, the standard statistical treatment
of 4-31G data gives[577] ΔH^o_{298} = -26.0 kJ/mol and $T\Delta S^o_{298}$ = -24.4 kJ/mol,
which means that the contributions to ΔG^o_{298} are very close in absolute
value. The resulting ΔG^o_{298} is -1.7 kJ/mol which is a value considerably different from ΔH^o_0 = -24.3 kJ/mol or even from ΔE_{SCF} = -34.4
kJ/mol.

With larger $(H_2O)_n$ clusters (for $n \geq 6$) nonregular structures become favored over cyclic ones. The binding energy per molecule is increasing with increasing cluster size; this indicates the stability of
the larger clusters against dissociation in any combination of smaller
ones. From this a very important conclusion may be drawn[595] viz. that
it does not seem to be possible to describe the energetic properties
of water as resulting from an ideal mixture of small regular clusters
as it is assumed in the "mixture" model of water. Moreover, the traditional quantum chemical approach to the determination of single optimum structure of a liquid is most likely not sufficiently meaningful
[584] because of a large number of configurations, almost equivalent in
energy, that must be statistically averaged anyway.

The importance of the many-body potentials in the expansion (5.38)
may be judged from Table 5.19. The entries of Table 5.19 show that
with the optimum ring structures $(H_2O)_3$ and $(H_2O)_4$ the nonadditivity

T a b l e 5.19

Two and higher body interactions in water trimer
and tetramer[595]

Potential	Energy [kJ/mol]	
	Trimer	Tetramer
$\sum_{i<j} V^{(2)}_{ij}$	-51.14	-85.91
$\sum_{i<j<k} V^{(3)}_{ijk}$	-4.73	-11.05
$V^{(4)}_{ijk\ell}$		-0.81
Total interaction energy	-55.87	-97.77

of pair interaction energies represents about 10% of the total interaction energy and that the $V^{(4)}$ contribution is negligible. The magnitude of $V^{(3)}$ depends strongly on geometry and in some cases it may
be repulsive[609]. Among interesting results of the examination of $V^{(3)}$
in the water trimer we note that[608,609] which confirms the concept of
cooperativity in H-bond formation. A comparable magnitude of $V^{(3)}$,
i.e. $\sim$10% of the total interaction energy, was also found[599] with
the complex H_2O-Li^+-H_2O. In this case, however, $V^{(3)}$ is repulsive, so
that the total interaction energy based on pair potentials only is
overestimated. Disregarding nonadditivity may give rise to the error
[611] as high as 18% (complexes $Al(H_2O)_n^{3+}$, n = 4-6), though $V^{(3)}$ may be
somewhat overestimated in this case owing to the basis set used. With
the asymmetric complex Li^+-H_2O-H_2O $V^{(3)}$ is negative. Its value[612]
given by a DZ+P basis set is -18.8 kJ/mol and, because ΣV_{ij} = -200.4
kJ/mol, it represents again roughly 10% of the total interaction energy. A high value of $V^{(3)}$ becomes particularly striking when it is
compared with $V^{(2)}_{H_2O-H_2O}$ = -17.1 kJ/mol given by the same basis set.
Combining[612] the two values gives us -35.9 kJ/mol for the strength of
the hydrogen bond. Hence, the presence of the lithium cation enhances
considerably H bonding. The same effect was also found with negative
ions, for example[613] with F^--H_2O-H_2O. From the two $Li^+(H_2O)_2$ complexes dealt with, the symmetrical one is more stable[614] than the a-
symmetric one. This might, of course, be expected because the symmetric complex conforms to the formation of the first solvation shell.
The second water molecule in the asymmetric complex may be viewed as
belonging to the second solvation shell, but the formation of the latter is only initiated first in $Li^+(H_2O)_5$. As we have learned, the $V^{(3)}$
potentials in the two complexes are of opposite sign and they have a
significant effect on the bonding[599,615]. The three-body potential may
be viewed as though it arises owing to the charge redistribution,
which is due to electrostatic polarization effects in the interaction
of the first two particles. Thorough investigation of charge redistribution in the two-particle interaction may be found, for example,
in papers by Schuster and co-workers[542,558,581]. However, also the exchange repulsion term is involved considerably in the nonadditivity
so that the origin of $V^{(3)}$ cannot be interpreted classically[609,615].

As regards the applications of the supermolecule approach to solvation of organic molecules, one can hardly afford to assume the whole
solvation shell. But it is still possible to arrive at meaningful results even for complex biomolecules by assuming only a few solvent
molecules. In the preceding section it was shown how the binding sites

of biomolecules may be determined. This permits selection and location of the most important molecules of the first solvation shell. Such a treatment was applied, for example, to the hydration of dimethyl-phosphate anion[616]. This approach cannot be expected to provide a complete description of the behavior of molecules in solution, but it should give an account of the main features of conformational changes[617] brought about by solvation.

In many cases useful information is obtained from calculations in which only one or two solvent molecules are assumed. An example of their application to electronic spectra is shown in Table 5.20. Not

T a b l e 5.20
Solvent effect on electronic transitions

Theoretical model	Approach	Transition	Solvent shift $[cm^{-1}]$	
			Calcd.	Exptl.[a]
Acetone-H_2O	STO-3G SCFCI	$n \rightarrow \pi^*$	1450[b]	1530
Acrolein-$(H_2O)_2$	4-31G EHP[c]	$n \rightarrow \pi^*$	4700[d]	2420[e]
Acrolein-$(H_2O)_2$	4-31G EHP[c]	$\pi \rightarrow \pi^*$	-2000[d]	-2280[e]

[a] For references see papers cited in footnotes b and d.
[b] Ref. 618.
[c] Electron-hole potential method.
[d] Ref. 619.
[e] Actually data for crotonaldehyde.

only do they reproduce qualitatively the observed solvent shifts but they also give us theoretical grounds for the observed facts (from the analysis of interaction energies and wave functions). One may proceed analogously in treatments on solvation phenomena observable by the infra-red spectroscopy, such as for example changes in vibrational modes of the solute owing to H bond formation[620] and of solvent molecules owing to the effect of ions[169,267,621,622].

Sometimes the problem of solvation may be reduced to the investigation of hydrogen-bonded systems of the type $(AH \cdots A)^{\pm}$, that represent an important entity in aqueous solutions of electrolytes. These systems possess double-well proton potential curves, if the A-A distance is larger[623] than about 2.4×10^{-10} m. Such a hydrogen bond is extremely polarizable[624] and owing to this high polarizability it interacts strongly with its environment[625,626]. With charged hydrogen

bonds, induced dipole interactions of the hydrogen bonds with the local electric fields from their environment are the most important. In
solution, the incidental local electric fields bring about various de-
formations of energy curves and this variability gives rise to continuous absorption in infra-red spectra. Details may be found in the
cited literature[624-626]. To demonstrate the application of the supermolecule approach to the environmental effect on hydrogen bonding, we
selected the results reported[627] for the system $Li^+-H_2O-OH^-$. These
are entered in Fig. 5.10, which shows the effects of the Li^+ ion and
hydration on the originally symmetrical proton potential curve for the
hydrogen bonding in H_2O-OH^-. It may be noticed that with the unsol-
vated system $Li^+-H_2O-OH^-$ a double well does not appear even at as large
O-O distance as 2.8×10^{-10} m but that the hydration tends to restore
the original symmetrical double-well character of the potential curve.
Energy curves for proton transfer are, of course, also met in many
other regions of chemistry. Extraordinarily important they are with
biochemical systems[208,628] where, however, the situation may be complicated by multiple hydrogen bonding[208,629]. Proton transfers are involved in many chemical problems and a detailed theoretical examination
of their potential curves may contribute to a deeper understanding of
various general problems of the chemical reactivity. An excellent example of the application of the supermolecule approach to the problem
of mechanism of the proton transfer is provided by the paper of Delpuech et al.[630]. From a comparison of potential curves for the proton
exchange in systems $NH_4^+-NH_3$, $NH_4^+-OH_2$, $NH_4^+-OH_2-OH_2$, $NH_4^+-OH_2-NH_3$, and
$H_2O-NH_4^+-OH_2-OH_2$ the cited authors were able to show that the process
of proton transfer between an ammonium ion, NH_4^+, and a water molecule
may take place only if assisted by solvation or by a concerted push-
-pull mechanism involving a third molecule, $NH_4^+ \rightarrow OH_2 \rightarrow NH_3$. Proton
transfer was also dramatically affected by the presence of water in
the system NH_3-HF. Obviously, in the absence of solvent the neutral
form is much more stable. When four or more H_2O molecules are applied,
however, the ionic form becomes more stable[631,632]. The same effect of
hydration is also seen in Fig. 5.10. The hydrogen $H_{(2)}$, which fluctu-
ates between $O_{(1)}$ and $O_{(2)}$, is shifted by hydration to the atom $O_{(1)}$,
which means that the ionized form $Li^+-H_2O-OH^-$ becomes favored.

Technical aspects of supermolecule calculation need not be noted
specifically since they are essentially the same as those mentioned in
the previous section. Hints for the basis set selection given in Table
5.16 may be supplemented with a note referring to proton potential
curves. Small basis sets give potential curves with a single minimum

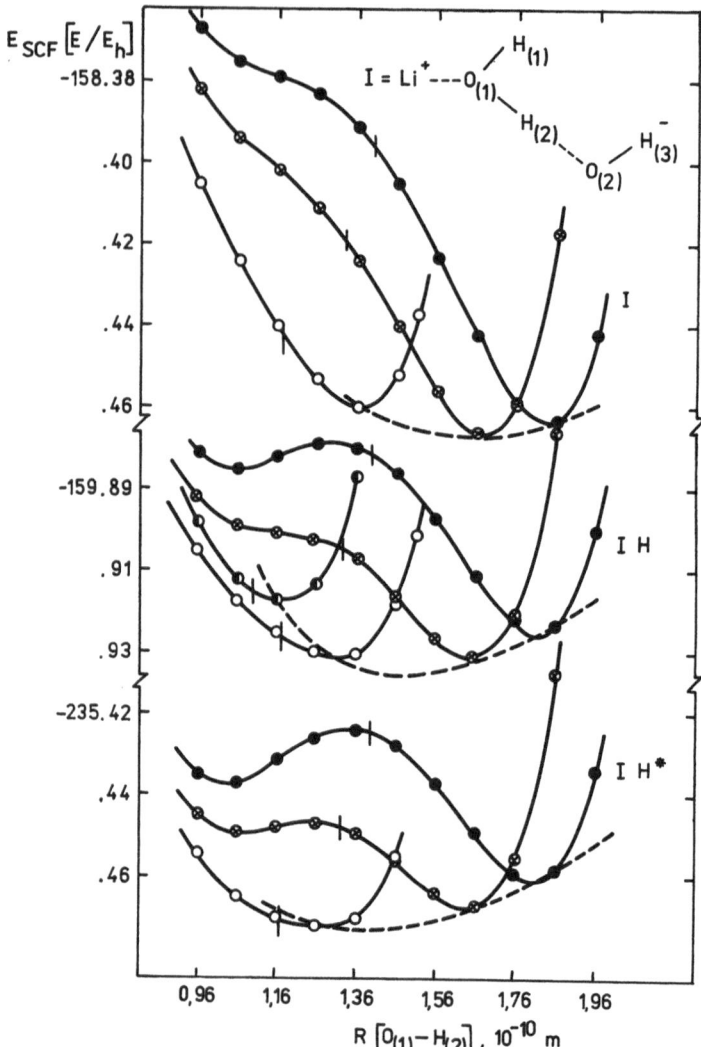

Figure 5.10

Proton potential curves[627] for the system $Li^+-H_2O-OH^-$ (at the top) and its hydrated models with three water molecules $(H_2O)_e$ around Li^+ in a tetrahedral coordination, the next two $(H_2O)_e$ molecules being coordinated to $H_{(1)}$ and $O_{(2)}$ (system IH in the middle). The subscript e in $(H_2O)_e$ means the representation of the water molecule by the electrostatic point charge model. The designation IH* (at the bottom) means a supermolecule calcula-tion in which only water molecules attached to Li^+ and $H_{(1)}$ were approximated by point charges. The dashed lines connect the minima of the curves for the assumed structures and they show the $O_{(1)}-O_{(2)}$ bond shortening owing to hydration. Verti-cal lines denote the location of $H_{(2)}$ at the midpoint of the $O_{(1)}-O_{(2)}$ bond. Bond lengths assumed for $O_{(1)}-O_{(2)}$ (in 10^{-10} m): ◐ 2.2; ○ 2.35; ⊗ 2.65; ● 2.8.

even for rather large distances between heavy atoms (e.g., O–O, N–N, O–N), for which large basis sets give[633] already a double well curve. Even when a double well nature of the potential curve is developed, small basis sets give a considerably lower barrier between the two minima. Correlation energy also reduces the barrier; it may bring about in some cases that the barrier disappears completely[588,634,635].

From a practical point of view the supermolecule approach has a considerable drawback viz. that it is very costly in terms of computer time for large supermolecules. This prompted attempts to incorporate in it economy-motivated simplifications. A common feature of all of them is the assumption about the electrostatic origin of the solute--solvent interactions, which, as we learned in Section 5.H., should be plausible with polar solvents. According to the idea of Ray[636] one simulates the solvent molecules by means of static dipoles, for which use is made of experimental values of dipole moments. The respective field integrals are easy to evaluate and they require very little additional computational effort over the SCF calculation on the solute molecule alone. Alternatively[637] one may represent the experimental dipole moment of the solvent molecule in the assumed position by a pair of point charges q_1 and q_2 with a proper separation and magnitude of q_1 and q_2. The dipole field approximation cannot of course be expected to give a quantitative account of solvation phenomena but merely qualitative predictions of trends. The applications to the NH_3–HF system showed that the gas-phase formation of NH_4F is facilitated[638] if the reactants are assumed as dimers $(NH_3)_2$ and $(HF)_2$ and that hydration has a stabilizing effect[637] on the hydrogen bond in NH_3–HF. Ordinarily, a somewhat more sophisticated point charge approximation is used[615,627,632,639–642], in which individual atoms of the solvent molecule are replaced by point charges. They are placed in the positions of atoms of the molecule to be approximated and their values are selected either on the basis of the Mulliken population analysis for a single solvent molecule or so that the dipole moment of the solvent molecule would be reproduced. Computer time saving is due to the fact that point charges contribute to the one-electron part of the Fock-operator only, in exactly the same way as they would if they were atomic nuclei. The two-electron integrals are identical with those for the free solute molecule. The only components of the interaction energy that are (approximately) accounted for in the point charge approximation are the Coulomb energy and a part of the polarization energy referring to the polarization of the solute molecule. Since the exchange repulsion energy is neglected, the distance between the solute molecule and

the approximated solvent molecules cannot be optimized. A mutual orientation may be optimized, however. The point charge approximation
was tested thoroughly by Noell and Morokuma[632]. Stated briefly, the
cited papers suggest that the results obtained with the point charge
approximation are in qualitative agreement with the predictions of
genuine supermolecule calculation. Quantitatively, the differences may
be rather large, however. Nevertheless, the utility of the point
charge model of solvation is indisputable: Noell and Morokuma[641], for
example, treated a model of the solvated Li^+ ion consisting of four,
eight and sixteen water molecules in the first, second and third solvation shell, respectively.

The last two approaches dealt with in this section are the continuum model[643-649] and closely related supermolecule-continuum model
[649-652]. The essence of the continuum model is that the system solute-
-solvent is assumed as a solute molecule inbedded in a polarizable dielectric continuum with a relative dielectric constant. Use is made of
classical theories of Born, Onsager and Kirkwood. Obviously, the model
is macroscopic in its nature and it disregards specific interactions
such as e.g., hydrogen bonds. Its use is therefore limited to solvents
with a low dielectric constant. Microscopic effects may be accounted
for by assuming a supermolecule composed of a solute molecule and a
few solvent molecules and by applying the continuum approach to the
supermolecule so formed. This is a principle of the supermolecule-continuum approach. Beveridge and Schnuelle[650,651] showed that the supermolecule-continuum approach may be very informative for the future development of theory of solutions. The idea of their approach is a
treatment of the solute molecule and the first solvation shell as a
supermolecule utilizing accurate quantum-mechanical representation of
all interactions in the supermolecular assembly, application of a continuum model to the interaction of this supermolecule with the polarizable dielectric and then configurational averaging by means of Monte
Carlo calculations. For the Gibbs free energy of hydration of Li^+,
Na^+, K^+, F^- and Cl^- ions, the present state of the agreement[651] between theory and experiment involves from 6 to 27% error, but the
theoretical approach used may certainly be refined.

In the continuum model, the solvent effect is accountable for in
two ways. Either one evaluates the solvation energy by means of explicit formulas derived in the classical theories noted above, or,
preferably, one may introduce the term for the solute-solvent interaction directly into the Hamiltonian[644-647]. The latter approach provides not only the solvation energy but also the wave function of the

solvated molecule, which permits applications to observables determined by the wave function (particularly applications in the field of
various spectroscopies are straightforward). A detailed discussion of
the existing continuum models is beyond the scope of the present book.
For their survey, comparison and applications we refer the reader to
a special monograph[519].

5.J. Presence and Future

This section may be considered as a sort of epilogue. We attempt
here to summarize briefly the present trends that permit deduction of
a further development in the near future. At present it is certainly
not necessary to advocate the utility of ab initio calculations, but
still it is useful to recall a tremendous progress made from the early
sixties. In the Introduction we quoted a paper by Allen and Karo[1] re-
viewing the ab initio calculations reported in the literature up to
1960. The largest among the systems treated was the 18-electron molecule SiH_4. Approximately a decade later it was possible to perform an
ab initio SCF calculation for a 136-electron complex cytosine-guanine
[208] with a basis set of 105 CGTF's and to compute even the potential
curve for the proton motion in the hydrogen bond of this system. It
must be admitted that it was an extraordinary and unique calculation
at that time, but calculations of a comparable scale followed soon. We
note on some of them to show that the time has already come when also
large chemical systems become amenable to ab initio applications.
Hence, Shipman and Christoffersen[162] reported calculations on a 144-
-electron polyglycine chain. In the same year, 1972, Clementi reported
[196] a calculation of the carbazole-trinitrofluorenone complex, a system consisting of 50 atoms and 232 electrons, with the basis set of
618 primitive GTF's grouped to 194 CGTF's. In 1973, Clementi and Popkie
[653] reported calculations with the basis set of 118 CGTF's on the barrier to internal rotation in the 158-electron complex sugar-phosphate-
-sugar, $C_{10}H_{19}O_8P$. A treatment of porphin with 136 CGTF's was reported by Almlöf[654] and recently a series of calculations was performed by Popkie, Kaufman and Koski on morphine, nalorphine[655], pro-
mazine and chlorpromazine[656]. The latter molecule possesses 40 nuclei
and 166 electrons and the basis set contained 132 CGTF's. Another type
of many-electron systems is represented by inorganic compounds and
organometalic complexes. We did not pay any special attention to them.
Actually ,they are amenable to treatment by any method dealt with in
this book, though it should be realized that mostly they possess so

many electrons that merely a minimum basis set can be applied. Nevertheless it is evident that the ab initio calculations are becoming a very useful tool for the rationalization and predictions of properties and reactivity of inorganic compounds. The present state in this field was reported by Veillard and Demuynck[657]. As an example of extremely extensive calculations on systems of this type we note a few papers. Bagus et al.[658] reported the treatment of the ferrocene molecule performed with the basis set of a somewhat better than the DZ quality containing 188 CGTF's. A basis set of 187 CGTF's was used in the calculations performed in Ohno's group[659,660]. The molecule treated was the cobalt-porphine complex containing 37 atoms and 187 electrons. The calculations of the largest nonpolymeric molecule treated to date were reported by Christoffersen[661]. This was the ethyl chlorophyllide _a_ molecule which contains 340 electrons.

It is obvious that SCF calculations for such large systems became feasible only by making use of all means noted in Chapter 3 that reduce the computer time. This concerns especially the integral part of computations. Although much progress has been made in this field, the problem of integral evaluation is still topical. It may be expected that a new development in this respect will be associated directly with the appearance of new programs. The point is that the algorithm improvement cannot be separated from computer programming. While with the majority of scientific calculations it is sufficient to make use of a standard level of a common programming language, very time-consuming ab initio calculations require to exploit the computer up to the very limits of its capacity. "In these situations a very careful understanding of the underlying principles of computer architecture is felt to be absolutely necessary, as well as a very extensive knowledge of the particular features and the system software of the computer actually to be used". This quotation is taken from a review by Diercksen and Kraemer[662] where the reader may find useful hints for computer programming.

Next, the development of computer technology itself should be noted. As is promised by computer manufactures, the near future should bring a remarkable advance in the construction of computers. If this will turn true, it may be hoped that the two principal requirements for a widespread use of ab initio calculations will be met - a considerably lower cost and a greater availability of effective computers. For the present time it appears appropriate to make use of minicomputers. The advantages of minicomputers are the following[663]. As indicated by the name, the minicomputers are of small size. Since they require no air conditioning, their placing in a department laboratory is free of complications. The price of a minicomputer is relatively low so that even a small theoretical group can afford it for its own use. The operation of the minicomputer is easy so that no extra paid operator is needed. Also the maintance cost is low, i.e. the costs of service contract, electricity, cards, paper etc. The minicomputers are sufficiently fast for making ab initio calculations and they may be under operation without interruption for a long time. The only disadvantage of minicomputers is a relatively small storage capacity. This may be overcome, however, at least to a certain degree, by an ingenious programming. This was clearly demonstrated by Sparks[664] who succeeded to adapt the Gaussian 70 program to a minicomputer with 128 K bytes of memory. Another example we note here comes from Schaefer's laboratory, where a new CI-SD program was developed and designed especially for a minicomputer[665,666]. The reported benchmark results[667] for the water molecule were superior to any result given in Table 2.15.

On several occasions we mentioned in this book a principal bottleneck of the ab initio SCF calculations – the n^4 problem involved in the computation of two-electron integrals. We showed that with large systems (with many large internuclear distances), the increase of the number of integrals is not as prohibitive as n^4. There remains, however, the problem of what to do with compact many-electron systems such as inorganic molecules containing atoms with high atomic numbers. Here the number of integrals may be reduced considerably by means of pseudopotential methods. Although, strictly speaking, the methods of this type lack rigor, they represent the only practical approach to molecules with heavier atoms so that one may expect their further development and increased use in the near future. Their importance is also enhanced by the fact that applications of semiempirical MO methods in inorganic chemistry have up to now met limited success. The use of pseudopotential methods is by no means restricted to molecules with heavy atoms. They also may be applied to first-row atom molecules, though in that case the effect is smaller. In the literature several versions of these methods have been reported. For their comparison, references and a detailed analysis we refer the reader to the paper by Kahn and collaborators[668]. A common feature of all these methods is that they reduce the all-electron problem to a valence-electron problem only. Inner shell electrons are approximated by an effective or model Hamiltonian which consists of a local potential and a set of projection operators. The former approximates the core Coulomb and exchange potentials whereas the latter prevents the collapse of the vadence orbitals into the core. The parameters for this potential may be obtained from the treatments of atoms made previously. The potential so constructed is then inserted into the one-electron part of the Hamiltonian for valence electrons. In this way the number of two-electron integrals is substantially reduced while the additional computer time for the evaluation of new integrals with the effective core potentials is comparable to that required for one electron integrals[668]. In actual calculations use may be made of standard basis sets - as an example we note the incorporation of a model potential method into the 4-31G method[669]. Compared to full ab initio calculations, the computer time saving depends, of course, on the particular molecule treated and the basis set used. For example, with chlorofluoromethanes[670] it ranges from 3.5 to a factor of almost 10, saving of one order was also found[671] with FeH^+, whereas[672] with Ge_2 the factor was as high as 50. With molecules containing first-row atoms only the time saving is[673,674] about 50%. As regards the comparison of computed quantities (orbital energies, one-electron properties, potential energy curves, etc.) with full ab initio results, mostly a very good agreement within a few percent is reported, the differences being smaller than differences due to basis set choices[668,675]. However, test calculations revealed cases where less satisfactory results are obtained. Hence, poor valence-core separability brings about overestimation of the interaction energy of K^+ and Ca^{++} cations with water[676] (as much as 15%). In contrast, the interaction energies of Li^+ and Na^+ cations with water[676,677] and formaldehyde[677] are underestimated because the pseudopotential approach is incapable of accounting for the attractive polarization of core electrons[677]. Obviously, much is left to be examined, for example the basis set effect or the effect of correlation energy[678]. In any case, however, the experience accumulated up to now suggests that SCF pseudopotential approaches became an important methodical tool of the present MO theory. In the near future, one may expect much progress in their applications, particularly in the studies of relativistic effects. Importance of relativistic effects was noted in Section 5.G in the connection with inner-shell ionization potentials of systems containing heavy atoms. Typically, the relativistic contributions to ionization energies were estimated in the cited papers on the basis of atomic calculations. However, attempts at a quantitative interpretation of properties of heavy-atom inorganic molecules stimulated development of methods in which relativistic effects would be inherently accounted for. It appears that the complexity of the problem may be reduced just by the

use of pseudopotential methods[679-685]. The results of actual calculations suggest that the relativistic effects may be important with the
physical properties of molecules containing atoms with the atomic numbers higher, say, than 30. Consider for example[679] the bond lengths in
the series CH_4, SiH_4, GeH_4, SnH_4 and PbH_4. While with CH_4, SiH_4 and
GeH_4 they remain almost unaffected by the relativistic correction,
with SnH_4 R/a_0 is changed from 3.41 to 3.33 and with PbH_4 from 3.60 to
3.40. Relevant changes were also found with force constants. Obviously,
for better understanding of relativistic effects further calculations
on molecules are very necessary.

It is fair to make also a note on methods in which the rigor of
ab initio SCF calculations is abandoned more drastically. In these
methods the number of integrals is reduced by adopting the approximations based on the zero differential overlap assumption and/or by
making use of mixed basis sets, which means that a part of integrals
is calculated with a smaller basis set. Unfortunately, these procedures are not numerically controlled, i.e. the neglect of integrals
is not based on rigorous pretests of integrals against a particular
threshold. The SCF calculations of this type may therefore provide
energies below the Hartree-Fock limit. It is of course a matter of
taste, but we think that these methods are less promising for further
development.

In spite of a rather large size of Chapter 4, we consider it useful
to add here a few remarks on the calculations of correlation energy.
Unlike with the SCF calculations, there is a large variety of theoretical approaches in this field that differ in their very essence, so
that it is by no means straightforward to attempt an outlook for the
next developments. As regards the present state, the methods most
widely used in practical applications are CI-SD and CEPA and one may
also expect their widespread use in the future. CI-SD calculations,
however, will most likely be consistently combined with the procedures
such as the Langhoff-Davidson formula[283] that remove the incorrect N-
-electron dependence. Anyway, a continuous development in the CI method is remarkable. Besides the achievements already noted in Section
4.D., the unitary group approach of Paldus[686,687] appears very appealing. Its graphical formulation[688] simplifies greatly the CI treatment and permits including effectively higher than double excitations.
Results of preliminary applications[689,690] are very promising and it
is believed[689] that for large CI problems the computation times will
be by an order in magnitude lower than those reported using state-of-
-the-art CI techniques. Siegbahn[690] made use of the unitary group ap-

proach for the generalization of the direct CI method. Direct evaluation of the CI energy from molecular integrals, without the diagonalization of the CI matrix, was already noted in Section 4.D. as an important tool permitting to perform extensive CI calculations. The o-
riginal method by Roos[285a] was developed with the aim of conforming it
to CI treatments covering singly and doubly excited states with respect to a closed shell SCF single determinant wave function. Later the
method was extended to more general cases[285b,690-692]. The treatment
of integrals in this approach is completely analogous to that involved
in the third-order MB-RSPT (and also in higher orders for doubly excited configurations (see Section 4.K.)).

The present trend in calculations with correlated wave functions
is to include higher than double excitations. Feasibility of CEPA calculations and their success in chemical applications belong certainly
to factors which benefited development in this direction. Explicit inclusion of certain contributions due to quadruple excitations, viz.
those that are due to disconnected wave function clusters of double
excitations, becomes now free of complications also in MB-RSPT through
fourth order. It is therefore every reason to expect that, besides
CI-SD and CEPA, MB-RSPT will soon become a method commonly used in
chemical applications. A fourth order MB-RSPT approach outlined in
Section 4.J. disregards triple excitations, which, however, are hardly
amenable to any existing effective method. Another topical problem is
a possible extension of MB-RSPT, so that it would permit convenient
treatment of the correlation problem for the multiconfiguration refer-
ence state. This is difficult with MB-RSPT, but the problem is tractable[693] with the ordinary Rayleigh-Schrödinger perturbation theory.
Finally, the CPMET method (see Chapter 4) should be noted. Present effort for inclusion of higher than double excitations will certainly
result in an increased number of applications of CPMET.

From the formulation of the correlation problem given in Chapter 4
for different approaches, it follows that essentially in any method
one needs to have available transformed two-electron integrals over
MO's. It has been shown that the number of operations involved in this
transformation increases as $\sim n^5$, if the size of the basis set is n.
If one recalls the problem of the "mere" n^4 dependence in SCF calculations and the effort for its reducing (Chapter 3), it becomes evident
that the progress in the solution of the problem of integral transformation is of crucial importance for further development of MO theories. Evidently, a new approach, other than that noted in Section 4.K.,
is needed. A possible solution of the problem might be a new algorithm

reported by Beebe and Linderberg[694] which is claimed to achieve car-
rying out the integral transformation in a small fraction ($\sim 1/5$ –
$1/3$) of the time required to generate the integrals over basis set
functions. The other way of time saving is to avoid, partly or completely, the explicit integral transformation. This idea was used probably for the first time in the program for CEPA calculations[293]. Similarly, one may achieve higher efficiency of third-order MB-RSPT calculations by incorporating a part of the integral transformation into
the expression for the energy, in which case the latter is evaluated
directly[253] from two-electron integrals over AO's. The same trick may
also be applied[295,695] in the CI-SD approach. If this technique is combined with the direct CI method, it is possible to avoid both the explicit integral transformation and the CI matrix construction[695]. Explicit integral transformation is also completely avoided in Meyer's
Self-Consistent-Electron- Pair Theory[696,697] which is due to an efficient operator formalism for obtaining the necessary matrix elements.
The procedures noted in this paragraphs were developed only recently,
so that one may expect that their further development will bring yet
more computational gain.

Advances in the development of theoretical methods and computer
construction are indispensable for the growing feasibility of ab initio
calculations, but this alone does not guarantee a future widespread use
of ab initio calculations by chemists in solving their problems. What
is demanded by chemists is a high predictive power of theory in various
branches of chemistry. A classical example of how the ab initio calculations should meet the needs of chemists was provided as early as in
1967 by Clementi and Gayles[251,544,698] in their study on the complex
$NH_3 \cdot HCl$. The calculation of the potential hypersurfaces and a detailed
analysis of wave functions of both the complex and the dissociation
components showed that NH_4Cl may exist in the gas phase. For the first
time, the results of ab initio calculations were used for the evaluation of the equilibrium constant for a chemical reaction[251]. Predicted
equilibrium constants for the process $NH_3(g) + HCl(g) \rightleftharpoons NH_4Cl(g)$ at
different temperatures suggested the experimental conditions at which
the complex NH_4Cl might be observed. Indeed, very soon after this prediction NH_4Cl has been detected[252] using high-temperature mass spec-
trometry. A very useful information was also obtained from the exploi-
tation of the Mulliken population analysis. From this purely theoretical concept about the electron distribution in molecules it was
concluded[698] that $NH_3 \cdot HCl$ is a complex with a strong hydrogen bond
with a proton shared by adjacent heavy atoms so that it can only

partly be viewed as an ion pair $NH_4^+.Cl^-$. This theoretical prediction
was confirmed fully by the infrared spectrum of the complex isolated
in a nitrogen matrix[699]. Among later examples of the predictive power
of ab initio calculations, the case of the energy barrier in the process $H' + FH \rightarrow H'F + H$ is very impressive. As noted already in Section
5.F., this reaction was believed up to recently to possess a low activation energy, the typical estimates being 4-30 kJ/mole. In contrast,
the ab initio calculations predicted[455-457] an activation barrier in
the range 180-205 kJ/mole. But on the basis of experience achieved
with ab initio calculations it was possible to claim that the true
barrier cannot be less than, say, 165 kJ/mole. This stimulated the experimental reexamination[458] from which it followed that the observed
reaction rate is compatible with the ab initio results and that in
fact the ab initio calculations gave a realistic account of the
$H' + FH \rightarrow H'F + H$ potential hypersurface prior to experimental confirmation. In general, the problems of chemical reactivity in the gas
phase are an ideal subject for ab initio treatments. The present methods are, at least with small molecules, capable of giving predictions
of such an accuracy that they may be taken as a reliable source of information in situations where the experimental examination of the
problem is difficult or hardly feasible. More specifically, the theoretical data on the energy hypersurfaces, geometries of intermediates,
energies of excited states and physicochemical molecular properties
may have a direct use in chemical kinetics, photochemistry[700], chemistry of the higher atmosphere and astrophysics[701]. Considerable progress is also countinuously made in applications in which the effect
of the surrounding medium is taken into account. Let us note at least
two fields of extraordinary importance - catalysis[702] and the biochemical applications where even as complicated problems as e.g., the
solvation of DNA[606] are becoming amenable to ab initio treatments.

Appendix A: Atomic Units

In ab initio calculations, so-called "atomic units" are used.
Their choice was not arbitrary but it was governed by practical considerations. Consider for example electronic repulsion integrals that
have the dimension of energy. In order to express them in SI units,
i.e. in J or m^{-1}, it would be necessary to express the elementary
charge in C. Since the nature of ab initio calculations requires to
compute electronic integrals with the precision to eight or more significant digits, the elementary charge should be given with the precision to at least nine significant digits. Even if the elementary
charge were known with such an accuracy, this approach would be rather
impractical. A confusion that might occur is well documented by programs for semiempirical calculations. We have found that in programs
coming from different laboratories the conversion factors differ frequently even in the fifth significant digit. To ensure perfect compatibility of ab initio calculations, a system of units was adopted
which avoids use of any conversion factor. Note, for example, that
the elementary charge in eqn. (1.1) is assumed to be unity. Tradition-
ally the most important among these "atomic units", energy and length,
are called "hartree" and "bohr", respectively.

However, general acceptance of SI units makes the use of atomic
units in publications prohibitive. This does not entail a problem if
the results are to be presented for observable quantities. A more complicated situation arises with the presentation of data such as e.g.
total energies and correlation energies which should serve as a comparative material for further calculations. A convenient way of preserving atomic units for practical needs but to avoid their meaning
of genuine units was suggested[703] by the IUPAC Commission on Physico-
chemical Symbols, Terminology and Units. It has been recommended to
give to computed quantities relative values according to the following
general scheme: usual symbol = calculated quantity/X.

For example,

Avoid:	Use:	
E = -1.1336 a.u.	$E = -1.1336\ E_h$	(in text)
E = -1.1336 hartree	$E/E_h = -1.1336$	(in tables)
r = 1.401 a.u.	$r = 1.401\ a_o$	(in text)
r = 1.401 bohr	$r/a_o = 1.401$	(in tables)

The following table represents a recommended treatment[703] for a series of selected quantities

Calculated physical quantity	Usual symbol	X	Value[703,704] of X
Length, (bond length)	ℓ,(r)	a_o	$5.2917706 \times 10^{-11}$ m
Mass	m	m_e	9.109534×10^{-31} kg
Energy	E	E_h	$4.3598144 \times 10^{-18}$ J
Molar energy	E,E_m	E_h	2.625500×10^{6} J mol^{-1}
Charge	Q	e	$1.6021892 \times 10^{-19}$ C
Electric dipole moment	p,μ	ea_o	8.4784×10^{-30} Cm
Electric quadrupole moment	Θ	ea_o^2	4.4866×10^{-40} Cm^2
Electric field	E	$E_h e^{-1} a_o^{-1}$	5.1423×10^{11} Vm^{-1}
Electric field gradient	-V_{zz}	$E_h e^{-1} a_o^{-2}$	9.7174×10^{21} Vm^{-2}
Probability density	Ψ^2(x,y,z)	a_o^{-3}	6.7483×10^{30} m^{-3}

Appendix B: Most Common Computer Programs

The creation of a computer program for ab initio calculations is a very hard job. This is nicely documented by the words of two among the authors of the POLYATOM system[210]: "The evolution of POLYATOM has been a notable event indeed. The work spans 25 years. It involves more than 20 persons, many of whom have never met. It was done in four countries at more than eight institutions and on a wide variety of computers". From this it is obvious that anybody who is going to start his own ab initio calculations faces a problem of procuring a suitable program. The aim of this Appendix is to facilitate this step by commenting briefly on the most common computer programs.

In spite of a great effort required for a development of programs, a relatively large number of them has been reported in the literature. We restrict here ourselves to programs that are intended for general molecular calculations, that are used by a wide chemical community and that are obtainable from the Quantum Chemistry Program Exchange at the Indiana University, Bloomington, Indiana 47401. The choice and the characterization of programs is of course greatly affected by personal taste and knowledge.

Gaussian 70[212] and Gaussian 76[705]

Gaussian 70 is probably the most widely used program for ab initio calculations. It meets perfectly the needs of users who prefer applications to developments of new computational methods. The program is designed predominantly for routine SCF calculations. It is highly efficient provided that use is made of basis sets of the "shell structure" (see Section 2.E. and 3.C.). In such a case of "standard" calculations the input and output are very simple and the preparation of a run does not require more work than with semiempirical calculations of the CNDO and INDO type. Once the program was adjusted to a particular computer, its handling is also very simple. Compared to other programs Gaussian 70 is advantageous for performing calculations for a large series of molecular structures. Options are available that permit automatic geometry optimization and potential energy scanning. No use is made of molecular symmetry. The available IBM 360/370[212] and CDC CYBER 706[706] versions are restricted to sp basis sets. The IBM 360/370 version is designed to run in a 226 K byte partition. Sparks has shown[664] however, that Gaussian 70 may be modified to run on minicomputers with

128 K bytes of memory, maintaining the dimension of 70 x 70 arrays. The extended version[705] of the program for CDC CYBER computers, Gaussian 76, may also accommodate d-functions. As in Gaussian 70, certain standard basis sets such as e.g. STO-3G, 4-31G and 6-31G* are stored internally for easy use.

POLYATOM (Version 2)[209]

A high percentage of ab initio calculations reported in the literature was performed using this program. Its use is versatile and it is amenable to any treatment noted in this book. The whole system is written in FORTRAN IV. The only relevant system dependent subroutines (written in a machine language) are those for packing and unpacking integers into and from a machine word. POLYATOM accommodates spdf Gaussian basis sets. It is oriented towards SCF calculations but a detailed documentation[209] and a clarity with which the program is written makes POLYATOM easily comprehensible and accessible to modifications and additions according to the user's needs. Owing to its versatility and flexibility POLYATOM possesses necessarily some features that some users may dislike. We comment on some of them. An efficient use of POLYATOM requires at least a limited knowledge of the algorithm, structure of the files on tapes (disk) and other programming details. A card input is rather complicated (especially for beginners), which makes large sequences of calculations (e.g. geometry optimization) somewhat troublesome. Compared to modern programs, such as PHANTOM or HONDO, POLYATOM is relatively slow. It may be concluded that rather than to routine SCF calculations POLYATOM is more suited to special applications and experimentation with new computational techniques. The program is available in versions for GE-635 and CDC 6600 computers [209] and for the IBM 360 series computers[707].

IBMOL (Version 4)[207]

Roughly speaking this program is of the same type as POLYATOM. The version distributed by QCPE is written for the IBM 7090 computer.

PHANTOM[708]

This is an extensively modified version of POLYATOM/2 for CDC 6000 and 7000 computers. It is a highly optimized program and therefore it is largely system-dependent. Some subroutines are written in a machine language and even in FORTRAN subroutines wide use is made of facilities specific for CDC computers. PHANTOM is a very efficient program. When comparing it with POLYATOM, a factor of 10 in speed is given, though it depends evidently on the molecule treated. This speeding up is not only due to optimized programming but also due to incorporated new theoretical developments such as the use of symmetry in the construction of F-matrix elements (see Section 3.D.). The card input is essentially the same as with POLYATOM.

HONDO

This is a very efficient SCF program with a very fast integral evaluation. For basis sets containing d functions, two-electron integral evaluation is roughly five times faster[214] than it is with PHANTOM. Molecular symmetry is utilized[236] efficiently in both integral evaluation and SCF procedure. The input with the symmetry information is very elegant, consisting only of a point group symbol and the coordinates of three points which specify the symmetry frame. Profit from using basis sets with the shell structure is fully exploited [214], though also general basis sets may be used. Both IBM and CDC versions of HONDO 76 are available[709]. Further development of the HONDO system resulted in the HONDO 78 program (for CDC computers), in which the SCF program was extended to the analytic computation of the energy gradient[409] with respect to nuclear coordinates. HONDO 78 also contains procedures in which the computed gradient may be employed for geometry optimization and force constant matrix calculation.

References

1 L. C. Allen and A. M. Karo: Rev. Mod. Phys. 32, 275 (1960).
2 S. Fraga, K. M. S. Saxena and J. Karwowski: Hartree-Fock Atomic
   Data. University of Alberta, Edmonton 1975, Canada.
3 S. Fraga, J. Karwowski and K. M. S. Saxena: Handbook of Atomic
   Data. Elsevier, Amsterdam 1976.
4 W. Kołos: Kwantowe teorie w chemii i biologii. Zakład Narodowy
   imenia Ossolińskich, Wroclaw 1971.
5 C. Roetti and E. Clementi: J. Chem. Phys. 60, 4725 (1974).
6 T. H. Dunning, Jr., R. M. Pitzer and S. Aung: J. Chem. Phys. 57,
   5044 (1972).
7 R. P. Hosteny, R. R. Gilman, T. H. Dunning, Jr., A. Pipano and
   I. Shavitt: Chem. Phys. Letters 7, 325 (1970).
8 B. J. Rosenberg and I. Shavitt: J. Chem. Phys. 63, 2162 (1975).
9 R. G. Parr: Quantum Theory of Molecular Electronic Structure.
   Benjamin, New York 1963.
10 P. T. van Duijnen and D. B. Cook: Mol. Phys. 21, 475 (1971); 22,
   637 (1971).
11 C. T. Llaguno, S. K. Gupta and S. M. Rothstein: Int. J. Quantum
   Chem. 7, 819 (1973).
12 J. C. Browne and R. D. Poshusta: J. Chem. Phys. 36, 1933 (1962).
13 E. Steiner and B. C. Walsh: J. Chem. Soc. Faraday Trans. II 71,
   921, 927 (1975).
14 K. Ruedenberg, R. C. Raffenetti and R. D. Bardo: in Energy, Structure and Reactivity. Proceedings of the 1972 Boulder Summer Research Conference on Theoretical Chemistry (eds. D. W. Smith and
   W. B. McRae). Wiley, New York 1973.
15 T. Živković and Z. B. Maksić: J. Chem. Phys. 49, 3083 (1968).
16 G. A. van der Velde: Dissertation, Groningen 1974.
17 K. Singer: Proc. Roy. Soc. A258, 412 (1960).
18 S. F. Boys: Proc. Roy. Soc. A258, 402 (1960).
19 J. C. Slater: Phys. Rev. 36, 57 (1930).
20 C. Zener: Phys. Rev. 36, 51 (1930).
21 C. C. J. Roothaan and P. S. Bagus: in Methods in Computational
   Physics, Vol. 2 (eds. B. Alder, S. Fernbach and M. Rotenberg).
   Academic Press, New York 1963.
22 E. Clementi and D. L. Raimondi: J. Chem. Phys. 38, 2686 (1963).
23 C. W. Bauschlicher, Jr. and H. F. Schaefer III: Chem. Phys.
   Letters 24, 412 (1974).

24 P. E. Cade, K. D. Sales and A. C. Wahl: J. Chem. Phys. 44, 1973 (1966).

25 W. M. Huo: J. Chem. Phys. 43, 624 (1965).

26 P. E. Cade and W. M. Huo: J. Chem. Phys. 47, 614 (1967).

27 P. E. Cade and W. M. Huo: J. Chem. Phys. 47, 649 (1967).

28 A. D. McLean and M. Yoshimine: Tables of Linear Molecule Wave Functions, Suppl. IBM J. Res. and Dev. 1967.

29 R. M. Stevens: J. Chem. Phys. 52, 1397 (1970).

30 P. S. Bagus, T. L. Gilbert and C. C. J. Roothaan: J. Chem. Phys. 56, 5195 (1972).

31 S. F. Boys: Proc. Roy. Soc. A200, 542 (1950).

32 S. Huzinaga, P. Palting and H. I. Flower: Phys. Rev. 6A, 2061 (1972); S. Huzinaga: Approximate Atomic Functions. III. University of Alberta, Edmonton, Canada 1973.

33 R. McWeeny: Nature 166, 21 (1950); A. Meckler: J. Chem. Phys. 21, 1750 (1953).

34 G. E. Kimball and G. F. Neumark: J. Chem. Phys. 26, 1285 (1957).

35 R. K. Nesbet: J. Chem. Phys. 32, 1114 (1960).

36 L. C. Allen: J. Chem. Phys. 37, 200 (1962).

37 M. Krauss: J. Chem. Phys. 38, 564 (1963).

38 S. Huzinaga: J. Chem. Phys. 42, 1293 (1965).

39 M. E. Schwartz and L. J. Schaad: J. Chem. Phys. 46, 4112 (1967).

40 I. G. Csizmadia, R. E. Kari, J. C. Polanyi, A. C. Roach and M. A. Robb: J. Chem. Phys. 52, 6205 (1970).

41 W. Kołos and C. C. J. Roothaan: Rev. Mod. Phys. 32, 219 (1960).

42 S. Fraga and B. J. Ransil: J. Chem. Phys. 35, 1967 (1961).

43 C. M. Reeves: J. Chem. Phys. 39, 1 (1963).

44 C. M. Reeves and M. C. Harrison: J. Chem. Phys. 39, 11 (1963).

45 I. G. Csizmadia, M. C. Harrison, J. W. Moskowitz and B. T. Sutcliffe: Theoret. Chim. Acta 6, 191 (1966).

46 K. O-ohata, H. Taketa and S. Huzinaga: J. Phys. Soc. Japan 21, 2306 (1966).

47 I. Shavitt: in Methods in Computational Physics, Vol. 2. (eds. B. Alder, S. Fernbach and M. Rotenberg). Academic Press, New York 1963.

48 E. Clementi and D. R. Davis: J. Computational Phys. 1, 223 (1966).

49 J. L. Whitten: J. Chem. Phys. 44, 359 (1966).

50 H. Taketa, S. Huzinaga and K. O-ohata: J. Phys. Soc. Japan 21, 2313 (1966).

51 D. R. Whitman and C. J. Hornback: J. Chem. Phys. 51, 398 (1969).

52 E. Clementi: Chem. Rev. 68, 341 (1968).

53 C. Salez and A. Veillard: Theoret. Chim. Acta 11, 441 (1968).

54 J. M. Schulman, J. W. Moskowitz and C. Hollister: J. Chem. Phys. 46, 2759 (1967).

55 C. D. Ritchie and H. F. King: J. Chem. Phys. 47, 564 (1967).

56 H. Basch, M. B. Robin and N. A. Kuebler: J. Chem. Phys. 47, 1201 (1967).

57 D. Neumann and J. W. Moskowitz: J. Chem. Phys. 49, 2056 (1968).

58 S. D. Peyerimhoff, R. J. Buenker and J. L. Whitten: J. Chem. Phys. 46, 1707 (1967).

59 T. H. Dunning, Jr.: J. Chem. Phys. 53, 2823 (1970).

60 T. H. Dunning, Jr.: Chem. Phys. Letters 7, 423 (1970).

61 D. J. David: Theoret. Chim. Acta 23, 226 (1971).

62 T. H. Dunning, Jr.: J. Chem. Phys. 55, 716 (1971).

63 R. C. Raffenetti: J. Chem. Phys. 58, 4452 (1973).

64 S. Huzinaga and C. Arnau: J. Chem. Phys. 52, 2224 (1970).

65 S. Huzinaga and Y. Sakai: J. Chem. Phys. 50, 1371 (1969).

66 S. Huzinaga: Approximate Atomic Functions. I, II. University of Alberta, Edmonton, Canada 1971.

67 F. B. van Duijneveldt: IBM, Tech. Res. Rep. RJ945 (1971).

68 P. G. Mezey, R. E. Kari and I. G. Csizmadia: J. Chem. Phys. 66, 964 (1977).

69 B. Roos and P. Siegbahn: Theoret. Chim. Acta 17, 209 (1970).

70 E. Clementi: J. Chem. Phys. 46, 4737 (1967).

71 S. Huzinaga and C. Arnau: J. Chem. Phys. 53, 348 (1970).

72 A. Veillard: Theoret. Chim. Acta 12, 405 (1968).

73 S. Rothenberg, R. H. Young and H. F. Schaefer III: J. Amer. Chem. Soc. 92, 3243 (1970).

74 H. Basch, C. J. Hornback and J. W. Moskowitz: J. Chem. Phys. 57, 1311 (1969).

75 A. J. H. Wachters: J. Chem. Phys. 52, 1033 (1970).

76 B. Roos, A. Veillard and G. Vinot: Theoret. Chim. Acta 20, 1 (1971).

77 T. H. Dunning, Jr.: J. Chem. Phys. 66, 1382 (1977).

78 S. Huzinaga: J. Chem. Phys. 66, 4245 (1977); 71, 1980 (1979).

79 J. M. Foster and S. F. Boys: Rev. Mod. Phys. 32, 303 (1960).

80 S. F. Boys and I. Shavitt: Proc. Roy. Soc. 254A, 487 (1960).

81 C. M. Reeves and R. Fletcher: J. Chem. Phys. 42, 4073 (1965).

82 M. Klessinger: Theoret. Chim. Acta 15, 353 (1969).

83 W. J. Hehre, R. F. Stewart and J. A. Pople: Symp. Faraday Soc. 2, 15 (1968).

84 W. J. Hehre, R. F. Stewart and J. A. Pople: J. Chem. Phys. 51, 2657 (1969).

85 W. J. Hehre, R. Ditchfield, R. F. Stewart and J. A. Pople: J. Chem. Phys. 52, 2769 (1970).

86 R. F. Stewart: J. Chem. Phys. 52, 431 (1970).

87 S. A. Clough, Y. Beers, G. P. Klein and L. S. Rothman: J. Chem. Phys. 59, 2254 (1973); T. R. Dyke and J. S. Muenter: J. Chem. Phys. 59, 3125 (1973).

88 W. A. Lathan, W. J. Hehre, L. A. Curtiss and J. A. Pople: J. Amer. Chem. Soc. 93, 6377 (1971).

89 J. E. Del Bene and J. A. Pople: J. Chem. Phys. 58, 3605 (1973).

90 M. Urban and R. Polák: Collect. Czech. Chem. Commun. 39, 2567 (1974).

91 R. Ditchfield, W. J. Hehre and J. A. Pople: J. Chem. Phys. 54, 724 (1971).

92 R. M. Pitzer and D. P. Merrifield: J. Chem. Phys. 52, 4782 (1970).

93 W. J. Hehre and J. A. Pople: J. Chem. Phys. 56, 4233 (1972).

94 W. J. Hehre and W. A. Lathan: J. Chem. Phys. 56, 5255 (1972).

95 W. J. Hehre, R. Ditchfield and J. A. Pople: J. Chem. Phys. 56, 2257 (1972).

96 R. Kari and I. G. Csizmadia: Int. J. Quantum Chem. 11, 441 (1977).

97 A. Rauk, L. C. Allen and E. Clementi: J. Chem. Phys. 52, 4133 (1970).

98 H. F. Schaefer III: in Critical Evaluation of Chemical and Physical Structural Information. National Academy of Sciences, Washington, D.C., 1974, p. 591.

99 H. F. Schaefer III: The Electronic Structure of Atoms and Molecules. A Survey of Rigorous Quantum Mechanical Results. Addison-Wesley, Reading, Massachusetts, 1972.

100 P. C. Hariharan and J. A. Pople: Mol. Phys. 27, 209 (1974).

101 J. B. Collins, P. v. R. Schleyer, J. S. Binkley and J. A. Pople: J. Chem. Phys. 64, 5142 (1976).

102 H. Lischka and V. Dyczmons: Chem. Phys. Letters 23, 167 (1973).

103 A. Veillard: Chem. Phys. Letters 4, 51 (1969).

104 J. P. Ranck and H. Johansen: Theoret. Chim. Acta 24, 334 (1972).

105 S. R. Ungemach and H. F. Schaefer III: Chem. Phys. Letters 38, 407 (1976).

106 R. Gleiter and A. Veillard: Chem. Phys. Letters 37, 33 (1976).

107 J. D. Dill and J. A. Pople: J. Chem. Phys. 62, 2921 (1975).

108 J. S. Binkley and J. A. Pople: J. Chem. Phys. 66, 879 (1977).

109 P. Čársky, I. Kozák, V. Kellö and M. Urban: Collect. Czech. Chem. Commun. 42, 1460 (1977).

110 I. H. Hillier and V. R. Saunders: Chem. Phys. Letters 4, 163 (1969).

111 B. Roos and P. Siegbahn: Theoret. Chim. Acta 17, 199 (1970).

112 U. Gelius, B. Roos and P. Siegbahn: Theoret. Chim. Acta 23, 59 (1971).

113 J. D. Petke and J. L. Whitten: J. Chem. Phys. 59, 4855 (1973).

114 T. H. Dunning, Jr.: J. Chem. Phys. 55, 3958 (1971).

115 F. P. Boer and W. N. Lipscomb: J. Chem. Phys. 50, 989 (1969).

116 I. Absar and J. R. van Wazer: Chem. Phys. Letters 11, 310 (1971).

117 P. C. Hariharan and J. A. Pople: Theoret. Chim. Acta 28, 213 (1973).

118 R. Ahlrichs, F. Driessler, H. Lischka, V. Staemmler and W. Kutzelnigg: J. Chem. Phys. 62, 1235 (1975).

119 R. Ahlrichs, F. Keil, H. Lischka, W. Kutzelnigg and V. Staemmler: J. Chem. Phys. 63, 455 (1975).

120 R. Ahlrichs, H. Lischka, B. Zurawski and W. Kutzelnigg: J. Chem. Phys. 63, 4685 (1975).

121 T. H. Dunning, Jr. and P. J. Hay: in Modern Theoretical Chemistry, Vol. 3 (ed. H. F. Schaefer III), Plenum Press, New York 1977.

122 M. Keeton and D. P. Santry: Chem. Phys. Letters 7, 105 (1970).

123 M. Urban, V. Kellö and P. Čársky: Theoret. Chim. Acta 45, 205 (1977).

124 H. Preuss: Z. Naturforsch. 11a, 823 (1956).

125 H. Preuss: Z. Naturforsch. 19a, 1335 (1964); 20a, 17, 21 (1965).

126 J. L. Whitten: J. Chem. Phys. 39, 349 (1963).

127 J. D. Petke, J. L. Whitten and A. W. Douglas: J. Chem. Phys. 51, 256 (1969).

128 S. Shih, R. J. Buenker, S. D. Peyerimhoff and B. Wirsam: Theoret. Chim. Acta 18, 277 (1970).

129 V. Staemmler: Theoret. Chim. Acta 35, 309 (1974).

130 F. Driessler and R. Ahlrichs: Chem. Phys. Letters 23, 571 (1973).

131 D. D. Shillady and F. S. Richardson: Chem. Phys. Letters 6, 359 (1970).

132 H. Le Rouzo and B. Silvi: Int. J. Quantum Chem. 13, 297, 311, 325 (1978).

133 S. Rothenberg and H. F. Schaefer III: J. Chem. Phys. 54, 2764 (1971).

134 W. Meyer and P. Pulay: J. Chem. Phys. 56, 2109 (1972).

135 P. Russegger, H. Lischka and P. Schuster: Chem. Phys. Letters 12, 392 (1971).

136 E. Glötzl: Dissertation, University of Vienna 1972.

137 S. D. Peyerimhoff and R. J. Buenker: Chem. Phys. Letters 16, 235 (1972).

138 T. Vladimiroff: Chem. Phys. Letters 24, 340 (1974).

139 W. Meyer and P. Pulay: Theoret. Chim. Acta 32, 253 (1974).

140 J. O. Jarvie, A. Rauk and C. Edmiston: Can. J. Chem. 52, 2778 (1974).

141 T. Vladimiroff: J. Chem. Phys. 64, 433 (1976).

142 P. G. Burton and B. R. Markey: Aust. J. Chem. 30, 231 (1977).

143 P. G. Burton: Mol. Phys. 34, 51 (1977).

144 N. R. Carlsen: Chem. Phys. Letters 51, 192 (1977).

145 M. Urban, S. Pavlík and T. Kožár: Chem. Zvesti 31, 165 (1977).

146 A. A. Frost: J. Chem. Phys. 47, 3707 (1967).

147 A. A. Frost, R. A. Rouse and L. Vescelius: Int. J. Quantum Chem. 2S, 43 (1968).

148 A. A. Frost: in Modern Theoretical Chemistry, Vol. 3 (ed. H. F. Schaefer III). Plenum Press, New York 1977.

149 P. H. Blustin and J. W. Linnett: J. Chem. Soc. Faraday Trans. II 70, 274 (1974).

150 E. R. Talaty, A. K. Schwartz and G. Simons: J. Amer. Chem. Soc. 97, 972 (1975).

151 A. A. Frost: J. Phys. Chem. 72, 1289 (1968).

152 L. P. Tan and J. W. Linnett: J. Chem. Soc. Chem. Commun. (1973) 736.

153 A. M. Semkow and J. W. Linnett: J. Chem. Soc. Faraday Trans. II 72, 1503 (1976).

154 R. E. Christoffersen: Advan. Quantum Chem. 6, 333 (1972).

155 J. L. Nelson and A. A. Frost: Theoret. Chim. Acta 29, 75 (1973).

156 P. H. Blustin and J. W. Linnett: J. Chem. Soc. Faraday Trans. II 70, 290 (1974).

157 B. Ford, G. G. Hall and J. C. Packer: Int. J. Quantum Chem. 4, 533 (1970).

158 R. M. Archibald, D. R. Armstrong and P. G. Perkins: J. Chem. Soc. Faraday Trans. II 70, 1557 (1974).

159 R. E. Christoffersen, D. Spangler, G. G. Hall and G. M. Maggiora: J. Amer. Chem. Soc. 95, 8526 (1973).

160 G. M. Maggiora and R. E. Christoffersen: J. Amer. Chem. Soc. 98, 8325 (1976).

161 R. E. Christoffersen: Int. J. Quantum Chem. 7S, 169 (1973).

162 L. L. Shipman and R. E. Christoffersen: Proc. Nat. Acad. Sci. USA 69, 3301 (1972).

163 R. K. Nesbet: J. Chem. Phys. 40, 3619 (1964).

164 D. P. Chong, F. G. Herring and D. McWilliams: J. Chem. Phys. 61, 78 (1974).

165 A. D. Buckingham: Advan. Chem. Phys. 12, 107 (1967).

166 D. E. Stogryn and A. P. Stogryn: Mol. Phys. 11, 371 (1966).

167 B. J. Rosenberg, W. C. Ermler and I. Shavitt: J. Chem. Phys. 65, 4072 (1976).

168 E. Clementi and H. Popkie: J. Chem. Phys. 57, 1077 (1972).

169 G. H. F. Diercksen, W. P. Kraemer and B. O. Roos: Theoret. Chim. Acta 36, 249 (1975).

170 P. Hennig, W. P. Kraemer, G. H. F. Diercksen and G. Strey: Theoret. Chim. Acta 47, 233 (1978).

171 R. S. Mulliken: J. Chem. Phys. 36, 3428 (1962).

172 F. Driessler, R. Ahlrichs, V. Staemmler and W. Kutzelnigg: Theoret. Chim. Acta 30, 315 (1973).

173 A. J. Duke and R. F. W. Bader: Chem. Phys. Letters 10, 631 (1971); R. F. W. Bader, A. J. Duke and R. R. Messer: J. Amer. Chem. Soc. 95, 7715 (1973).

174 V. Dyczmons and W. Kutzelnigg: Theoret. Chim. Acta 33, 239 (1974).

175 F. Keil and R. Ahlrichs: J. Amer. Chem. Soc. 98, 4787 (1976).

176 M. M. Heaton: J. Chem. Phys. 67, 2925 (1977).

177 J. Simons: Ann. Rev. Phys. Chem. 28, 15 (1977).

178 P. A. Benioff: Theoret. Chim. Acta 48, 337 (1978).

179 H. Lischka, P. Čársky and R. Zahradník: Chem. Phys. 25, 19 (1977).

180 P. Čársky, R. Zahradník, M. Urban and V. Kellö: Chem. Phys. Letters 61, 85 (1979); Collect. Czech. Chem. Commun. 43, 1965 (1978).

181 P. J. Hay: J. Chem. Phys. 66, 4377 (1977).

182 J. E. Mentall, E. P. Gentieu, M. Krauss and D. Neumann: J. Chem. Phys. 55, 5471 (1971).

183 W. Coughran, J. Rose, T. I. Shibuya and V. McKoy: J. Chem. Phys. 58, 2699 (1973).

184 U. Fischbach, R. J. Buenker and S. D. Peyerimhoff: Chem. Phys. 5, 265 (1974).

185 D. Demoulin: Chem. Phys. 17, 471 (1976).

186 R. J. Buenker and S. D. Peyerimhoff: Chem. Phys. Letters 36, 415 (1975).

187 I. Easson and M. H. L. Pryce: Can. J. Phys. 51, 518 (1973).

188 J. A. Hall and W. G. Richards: Mol. Phys. 23, 331 (1972).

189 R. J. Buenker and S. D. Peyerimhoff: Chem. Phys. 9, 75 (1976).

190 R. J. Buenker, S. Shih and S. D. Peyerimhoff: Chem. Phys. Letters 44, 385 (1976).

191 H. J. Werner and W. Meyer: Mol. Phys. 31, 855 (1976).

192 P. Swanstrøm, W. P. Kraemer and G. H. F. Diercksen: Theoret.
     Chim. Acta 44, 109 (1977).
193 A. J. Sadlej: Mol. Phys. 34, 731 (1977).
194 A. J. Sadlej: Theoret. Chim. Acta 47, 205 (1978); K. Woliński
     and A. J. Sadlej: Chem. Phys. Letters 64, 51 (1979); A. J. Sadlej:
     J. Phys. Chem. 83, 1653 (1979).
195 C. C. J. Roothaan: Rev. Mod. Phys. 23, 69 (1951).
196 E. Clementi: Proc. Nat. Acad. Sci. USA 69, 2942 (1972).
197 E. Clementi: in Selected Topics in Molecular Physics (ed. E.
     Clementi). Verlag Chemie, Weinheim 1972.
198 V. R. Saunders: in Computational Techniques in Quantum Chemistry
     and Molecular Physics (eds. G. H. F. Diercksen, B. T. Sutcliffe
     and A. Veillard). D. Reidel, Dordrecht 1975.
199 R. G. Parr and B. L. Crawford, Jr.: Proc. Nat. Acad. Sci. USA 38,
     547 (1952).
200 F. J. Corbato: J. Chem. Phys. 24, 452 (1956).
201 C. C. J. Roothaan: J. Chem. Phys. 28, 982 (1958).
202 I. Shavitt and M. Karplus: J. Chem. Phys. 36, 550 (1962).
203 S. F. Boys, G. B. Cook, C. M. Reeves and I. Shavitt: Nature 178,
     1207 (1956).
204 I. Shavitt and M. Karplus: J. Chem. Phys. 43, 398 (1965).
205 M. P. Barnett: in Methods in Computational Physics (eds. B. Alder,
     S. Fernbach and M. Rotenberg),Vol. 2. Academic Press, New York
     1963.
206 W. E. Palke and W. N. Lipscomb: J. Amer. Chem. Soc. 88, 2384
     (1966).
207 A. Veillard: IMBOL-4, Computation of Wave Functions for Molecules
     of General Geometry, Version 4. IBM Res. Laboratory, San Jose
     1968.
208 E. Clementi, J. Mehl and W. von Niessen: J. Chem. Phys. 54, 508
     (1971).
209 D. B. Neumann, H. Basch, R. L. Kornegay, L. S. Snyder, J. W.
     Moskowitz, C. Hornback and S. P. Liebmann, POLYATOM (Version 2).
     QCPE 199, Indiana University, Bloomington.
210 J. W. Moskowitz and L. C. Snyder: in Modern Theoretical Chemistry,
     Vol. 3 (ed. H. F. Schaefer III). Plenum Press, New York 1977.
211 R. Ahlrichs: Theoret. Chim. Acta 33, 157 (1974).
212 W. J. Hehre, W. A. Lathan, R. Ditchfield, M. D. Newton and J. A.
     Pople, GAUSSIAN 70, QCPE 236. Indiana University, Bloomington.
213 J. A. Pople and W. J. Hehre: J. Computational Phys. 27, 161 (1978).
214 M. Dupuis, J. Rys and H. F. King: J. Chem. Phys. 65, 111 (1976).

215 V. Dyczmons: Theoret. Chim. Acta 28, 307 (1973).

216 M. Urban and P. Hobza: Theoret. Chim. Acta 36, 207 (1975).

217 J. L. Whitten: J. Chem. Phys. 58, 4496 (1973).

218 D. L. Wilhite and R. N. Euwema: J. Chem. Phys. 61, 375 (1974).

219 A. Veillard: in Computational Techniques in Quantum Chemistry
     and Molecular Physics (eds. G. H. F. Diercksen, B. T. Sutcliffe
     and A. Veillard). D. Reidel, Dordrecht 1975.

220 J. H. Letcher, I. Absar and J. R. Van Wazer: Int. J. Quantum
     Chem. S6, 451 (1972).

221 J. H. Letcher: J. Chem. Phys. 54, 3215 (1971).

222 L. L. Shipman and R. E. Christoffersen: Chem. Phys. Letters 15,
     469 (1972).

223 P. de Montgolfier and A. Hoareau: J. Chem. Phys. 65, 2477 (1976).

224 M. F. Guest and V. R. Saunders: Mol. Phys. 28, 819 (1974).

225 C. C. J. Roothaan: Rev. Mod. Phys. 32, 179 (1960).

226 V. R. Saunders and I. H. Hillier: Int. J. Quantum Chem. 7, 699
     (1973).

227 R. C. Raffenetti: Chem. Phys. Letters 20, 335 (1973).

228 P. Siegbahn: Chem. Phys. Letters 8, 245 (1971).

229 H. Preuss and G. Diercksen: Int. J. Quantum Chem. 1, 605 (1967).

230 P. D. Dacre: Chem. Phys. Letters 7, 47 (1970).

231 N. W. Winter, W. C. Ermler and R. M. Pitzer: Chem. Phys. Letters
     19, 179 (1973).

232 D. F. Brailsford and J. Hylton: Chem. Phys. Letters 18, 595 (1973).

233 M. Elder: Int. J. Quantum Chem. 7, 75 (1973).

234 R. M. Pitzer: J. Chem. Phys. 58, 3111 (1973).

235 P. S. Bagus and U. I. Wahlgren: Computers & Chemistry 1, 95 (1976).

236 M. Dupuis and H. F. King: Int. J. Quantum Chem. 11, 613 (1977).

237 P. Čársky and I. Hubač: Chem. listy 71, 673 (1977) (in Czech).

238 I. Hubač and P. Čársky: Topics Curr. Chem. 75, 97 (1978).

239 R. K. Nesbet: Advan. Chem. Phys. 9, 321 (1965).

240 W. Kutzelnigg: Topics Curr. Chem. 41, 31 (1973).

241 L. C. Snyder: J. Chem. Phys. 46, 3602 (1967).

242 L. C. Snyder and H. Basch: J. Amer. Chem. Soc. 91, 2189 (1969).

243 P. George, M. Trachtman, A. M. Brett and C. W. Bock: Int. J.
     Quantum Chem. 12, 61 (1977).

244 P. George, M. Trachtman, A. M. Brett and C. W. Bock: J. Chem.
     Soc., Perkin Trans. 2, 1036 (1977).

245 W. J. Hehre, R. T. McIver, Jr., J. A. Pople and P. v. R. Schleyer:
     J. Amer. Chem. Soc. 96, 7162 (1974).

246 L. Radom: J. Chem. Soc., Chem. Comm., (1974) 403.

247 W. J. Hehre, D. Ditchfield, L. Radom and J. A. Pople: J. Amer. Chem. Soc. 92, 4796 (1970).

248 L. Radom, W. J. Hehre and J. A. Pople: J. Chem. Soc., Inorg. Phys. Theor. (1971) 2299.

249 L. Radom, W. J. Hehre and J. A. Pople: J. Amer. Chem. Soc. 93, 289 (1971).

250 R. Ahlrichs: Theoret. Chim. Acta 35, 59 (1974).

251 E. Clementi and J. N. Gayles: J. Chem. Phys. 47, 3837 (1967).

252 P. Goldfinger and G. Verhaegen: J. Chem. Phys. 50, 1467 (1969).

253 J.A. Pople, J. S. Binkley and R. Seeger: Int. J. Quantum Chem. S10,1 (1976).

254 R. F. W. Bader and R. A. Gangi: in Theoretical Chemistry, Vol. 2 (eds. R. N. Dixon and C. Thomson). The Chemical Society, London 1975.

255 G. C. Lie and E. Clementi: J. Chem. Phys. 60, 1288 (1974).

256 G. Das and A. C. Wahl: J. Chem. Phys. 44, 87 (1966).

257 R. L. Matcha: J. Chem. Phys. 48, 335 (1968).

258 A. C. Hurley: Advan. Quant. Chem. 7, 315 (1973).

259 J. B. Moffat: J. Mol. Structure 15, 325 (1973).

260 H. Ö. Pamuk: Theoret. Chim. Acta 28, 85 (1972).

261 O. Sinanoğlu and H. Ö. Pamuk: J. Amer. Chem. Soc. 95, 5435 (1973).

262 E. Clementi and H. Popkie: J. Chem. Phys. 57, 4870 (1972).

263 G. C. Lie and E. Clementi: J. Chem. Phys. 60, 1275 (1974).

264 R. Colle and O. Salvetti: Theoret. Chim. Acta 37, 329 (1975).

265 E. Wigner: Phys. Rev. 46, 1002 (1934).

266 H. Kistenmacher, H. Popkie, E. Clementi and R. O. Watts: J. Chem. Phys. 60, 4455 (1974).

267 H. Kistenmacher, H. Popkie and E. Clementi: J. Chem. Phys. 59, 5842 (1973).

268 O. Sinanoğlu and H. Ö. Pamuk: Theoret. Chim. Acta 27, 289 (1972).

269 H. Ö. Pamuk: J. Amer. Chem. Soc. 98, 7948 (1976); H. Ö. Pamuk and C. Trindle: Int. J. Quantum Chem. S12, 271 (1978).

270 P. Čársky, R. Zahradník and P. Hobza: Theoret. Chim. Acta 40, 287 (1975).

271 J. Koller, T. Šolmajer, B. Borštnik and A. Ažman: J. Mol. Structure 26, 439 (1975).

272 P. Hobza, P. Čársky and R. Zahradník: Collect. Czech. Chem. Commun. 43, 676 (1978).

273 G. Verhaegen, W. G. Richards and C. M. Moser: J. Chem. Phys. 47, 2595 (1967).

274 H. P. D. Liu and G. Verhaegen: J. Chem. Phys. 53, 735 (1970).

275 I. Shavitt: in Modern Theoretical Chemistry, Vol. 3 (ed. H. F. Schaefer III). Plenum Press, New York 1977.

276 E. R. Davidson: in The World of Quantum Chemistry. Proceedings of the First International Congress on Quantum Chemistry (eds. R. Daudel and B. Pullman). D. Reidel, Dordrecht 1974.

277 P. S. Bagus, B. Liu, A. D. McLean, M. Yoshimine: in Computational Methods for Large Molecules and Localized States in Solids (eds. F. Herman, A. D. McLean and R. K. Nesbet). Plenum Press, New York 1973.

278 R. J. Buenker and S. D. Peyerimhoff: Theoret. Chim. Acta 35, 33 (1974).

279 B. Roos: in Computational Techniques in Quantum Chemistry and Molecular Physics (eds. G. H. F. Diercksen, B. T. Sutcliffe and A. Veillard). D. Reidel, Dordrecht 1975.

280 A. C. Hurley: Electron Correlation in Small Molecules. Academic Press, London 1977.

281 F. Grimaldi, A. Lecourt and C. Moser: Int. J. Quantum Chem. S1, 153 (1967).

282 C. F. Bunge: Phys. Rev. 168, 92 (1968).

283 S. R. Langhoff and E. R. Davidson: Int. J. Quantum Chem. 8, 61 (1974).

284 M. Yoshimine: J. Computational Phys. 11, 449 (1973).

285a B. Roos: Chem. Phys. Letters 15, 153 (1972).

285b B. O. Roos and P. E. M. Siegbahn: in Modern Theoretical Chemistry, Vol. 3 (ed. H. F. Schaefer III). Plenum Press, New York 1977.

286 R. J. Buenker and S. D. Peyerimhoff: Theoret. Chim. Acta 39, 217 (1975).

287 R. Ahlrichs and W. Kutzelnigg: J. Chem. Phys. 48, 1819 (1968).

288 W. Meyer: in Modern Theoretical Chemistry, Vol. 3 (ed. H. F. Schaefer III). Plenum Press, New York 1977.

289 C. F. Bender and E. R. Davidson: J. Phys. Chem. 70, 2675 (1966).

290 H. F. Schaefer III and C. F. Bender: J. Chem. Phys. 55, 1720 (1971).

291 C. F. Bender and H. F. Schaefer III: J. Chem. Phys. 55, 4798 (1971).

292 W. Meyer: Int. J. Quantum Chem. S5, 341 (1971).

293 R. Ahlrichs, H. Lischka, V. Staemmler and W. Kutzelnigg: J. Chem. Phys. 62, 1225 (1975).

294 R. J.Bartlett and I. Shavitt: Int. J. Quantum Chem. S11, 165 (1977); S12, 543 (1978).

295 J. A. Pople, R. Seeger and R. Krishnan: Int. J. Quantum Chem. S11, 149 (1977).

296 P. E. M. Siegbahn: Chem. Phys. Letters 55, 386 (1978).

297 W. L. Luken: Chem. Phys. Letters 58, 421 (1978).

298 R. Ahlrichs: Comput. Phys. Comm. 17, 31 (1979).

299 A. C. Wahl and G. Das: Advan. Quantum Chem. 5, 261 (1970).

300 J. Hinze: J. Chem. Phys. 59, 6424 (1973).

301 A. Veillard: Theoret. Chim. Acta 4, 22 (1966); E. Clementi and A. Veillard. J. Chem. Phys. 44, 3050 (1966).

302 A. C. Wahl and G. Das: in Modern Theoretical Chemistry, Vol. 3 (ed. H. F. Schaefer III). Plenum Press, New York 1977.

303 A. R. Gregory: Chem. Phys. Letters 11, 271 (1971).

304 A. R. Gregory and M. N. Paddon-Row: Chem. Phys. Letters 12, 552 (1972).

305 O. Sinanoğlu: J. Chem. Phys. 36, 706 (1962).

306 R. K. Nesbet: Advan. Chem. Phys. 14, 1 (1969).

307 M. Jungen and R. Ahlrichs: Theoret. Chim. Acta 17, 339 (1970).

308 M. Gelus, R. Ahlrichs, V. Staemmler and W. Kutzelnigg: Theoret. Chim. Acta 21, 63 (1971).

309 M. Gelus and W. Kutzelnigg: Theoret. Chim. Acta 28, 103 (1973).

310 H. Lischka: Theoret. Chim. Acta 31, 39 (1973).

311 V. Staemmler: Theoret. Chim. Acta 31, 49 (1973).

312 W. Kutzelnigg, V. Staemmler and C. Hoheisel: Chem. Phys. 1, 27 (1973).

313 B. Tsapline and W. Kutzelnigg: Chem. Phys. Letters 23, 173 (1973).

314 H. Lischka: Chem. Phys. Letters 20, 448 (1973).

315 H. Lischka: Chem. Phys. 2, 191 (1973).

316 H. Lischka: J. Amer. Chem. Soc. 96, 4761 (1974).

317 J. Čížek: J. Chem. Phys. 45, 4256 (1966).

318 H. Primas: in Modern Quantum Chemistry. Istanbul Lectures (ed. O. Sinanoğlu). Academic Press, New York 1965.

319 W. Kutzelnigg: in Modern Theoretical Chemistry, Vol. 3 (ed. H. F. Schaefer III). Plenum Press, New York 1977.

320 J. Paldus and J. Čížek: Advan. Quant. Chem. 9, 105 (1975).

321 R. E. Watson: Phys. Rev. 119, 170 (1960).

322 W. Kutzelnigg: Methoden zur Berücksichtigung der Elektronen- korrelation. Ruhr-Universität, Bochum 1976.

323 J. Paldus, J. Čížek and I. Shavitt: Phys. Rev. A5, 50 (1972).

324 P. R. Taylor, G. B. Bacskay, N. S. Hush and A. C. Hurley: Chem. Phys. Letters 41, 444 (1976); J. Chem. Phys. 69, 1971, 4669 (1978); 70, 4481 (1979).

325 J. A. Pople, R. Krishnan, H. B. Schlegel and J. S. Binkley: Int. J. Quantum Chem. 14, 545 (1978).

326a R. J. Bartlett and G. D. Purvis: Int. J. Quantum Chem. 14, 561 (1978).

326b R. J. Bartlett, I. Shavitt and G. D. Purvis III: J. Chem. Phys. 71, 281 (1979).

327 A. Pipano and I. Shavitt: Int. J. Quantum Chem. 2, 741 (1968).

328 J. Paldus and J. Čížek: in Energy, Structure and Reactivity. Proceedings of the 1972 Boulder Summer Research Conference on Theoretical Chemistry (eds. D. W. Smith and W. B. McRae). Wiley, New York 1973.

329 W. Meyer: J. Chem. Phys. 58, 1017 (1973).

330 H. P. Kelly: Phys. Rev. A134, 1450 (1964).

331 H. P. Kelly: Phys. Rev. B136, 896 (1964).

332 W. Meyer: Theoret. Chim. Acta 35, 277 (1974).

333 W. Meyer and P. Rosmus: J. Chem. Phys. 63, 2356 (1975).

334 P. Rosmus and W. Meyer: J. Chem. Phys. 66, 13 (1977).

335 M. A. Robb: in Computational Techniques in Quantum Chemistry and Molecular Physics (eds. G. H. F. Diercksen, B. T. Sutcliffe and A. Veillard). D. Reidel, Dordrecht 1975.

336 R. J. Bartlett and D. M. Silver: Int. J. Quantum Chem. S9, 183 (1975).

337 R. Manne: Int. J. Quantum Chem. S11, 175 (1977).

338 C. Møller and M. S. Plesset: Phys. Rev. 46, 618 (1934).

339 J. Goldstone: Proc. Roy. Soc. A239, 267 (1957).

340 R. J. Bartlett and D. M. Silver: J. Chem. Phys. 62, 3258 (1975).

341 D. M. Silver and R. J. Bartlett: Phys. Rev. A13, 1 (1976).

342 R. J. Bartlett, S. Wilson and D. M. Silver: Int. J. Quantum Chem. 12, 737 (1977).

343 D. L. Freemen and M. Karplus: J. Chem. Phys. 64, 2641 (1976).

344 M. Urban, V. Kellö and I. Hubač: Chem. Phys. Letters 51, 170 (1977).

345 E. L. Mehler, G. A. van der Velde and W. C. Nieuwpoort: Int. J. Quantum Chem. S9, 245 (1975).

346 S. Wilson, D. M. Silver and R. A. Farrell: Proc. Roy. Soc. A356, 363 (1977).

347 R. J. Bartlett and I. Shavitt: Chem. Phys. Letters 50, 190 (1977).

348 S. Wilson: J. Phys. B 12, 1623 (1979).

349 L. S. Cederbaum, K. Schönhammer and W. v. Niessen: Chem. Phys. Letters 34, 392 (1975).

350 V. Kellö, M. Urban, I. Hubač and P. Čársky: Chem. Phys. Letters 58, 83 (1978).

351 S. Wilson: J. Chem. Phys. 67, 4491 (1977); Int. J. Quantum Chem. 12, 609 (1977); Mol. Phys. 35, 1 (1978).

352 M. Urban and V. Kellö: Mol. Phys. 38, 1621 (1979).

353 K. A. Brueckner and C. A. Levinson: Phys. Rev. 97, 1344 (1955).

354 J. Čížek: Advan. Chem. Phys. 14, 35 (1969).

355 V. Kvasnička and V. Laurinc: Theoret. Chim. Acta 45, 197 (1977).

356 R. J. Bartlett and D. M. Silver: in Quantum Science (eds. J. L. Calais, O. Goscinski, J. Linderberg and Y. Öhrn). Plenum Press, New York 1976.

357 H. P. Kelly: Phys. Rev. 144, 39 (1966).

358 M. Gell-Mann and K. A. Brueckner: Phys. Rev. 106, 364 (1957).

359 S. Wilson and D. M. Silver: Phys. Rev. 14, 1949 (1976).

360 N. S. Ostlund and M. F. Bowen: Theoret. Chim. Acta 40, 175 (1975).

361 P. Claverie, S. Diner and J. P. Malrieu: Int. J. Quantum Chem. 1, 751 (1967).

362 P. S. Epstein: Phys. Rev. 28, 695 (1926).

363 R. K. Nesbet: Proc. Roy. Soc. A230, 312 (1955).

364 G. D. Purvis and R. J. Bartlett: J. Chem. Phys. 68, 2114 (1978).

365 R. Krishnan and J. A. Pople: Int. J. Quantum Chem. 14, 91 (1978).

366 I. Hubač, M. Urban and V. Kellö: Chem. Phys. Letters 62, 584 (1979).

367 M. Urban, I. Hubač, V. Kellö and J. Noga: J. Chem. Phys. 72, 3378 (1980).

368a S. Wilson and D. M. Silver: Mol. Phys. 36, 1539 (1978).

368b S. Wilson and D. M. Silver: Int. J. Quantum Chem. 15, 683 (1979).

369 S. Wilson and V. R. Saunders: J. Phys. B 12, L403 (1979).

370a L. T. Redmon, G. D. Purvis and R. J. Bartlett: J. Chem. Phys. 69, 5386 (1978).

370b L. T. Redmon, G. D. Purvis III and R. J. Bartlett: J. Amer. Chem. Soc. 101, 2856 (1979).

371a A. Meunier, B. Lévy and G. Berthier: Int. J. Quantum Chem. 10, 1061 (1976).

371b W. Kutzelnigg, A. Meunier, B. Lévy and G. Berthier: Int. J. Quantum Chem. 12, 777 (1977).

372 B. H. Brandow: Rev. Mod. Phys. 39, 771 (1967).

373 I. Hubač: Int. J. Quantum Chem., in press.

374 J. Čížek and J. Paldus: Int. J. Quantum Chem. 5, 359 (1971).

375 B. T. Sutcliffe: in Proceedings of the Seminar on Computational Problems in Quantum Chemistry (eds. A. Veillard and G. H. F. Diercksen). Strasbourg 1969.

376   C. F. Bender: J. Computational Phys. 9, 547 (1972).

377   G. H. F. Diercksen: Theoret. Chim. Acta 33, 1 (1974).

378   P. Pendergast and W. H. Fink: J. Computational Phys. 14, 286
      (1974); P. Pendergast and E. F. Hayes: J. Computational Phys.
      26, 236 (1978).

379   H. Le Rouzo, G. Raseev and B. Silvi: Computers & Chemistry 2,
      15 (1978).

380   M. Urban and V. Kellö: unpublished.

381   D. M. Silver: Comput. Phys. Commun. 14, 71 (1978).

382   D. M. Silver: Comput. Phys. Commun. 14, 81 (1978).

383   S. Wilson: Comput. Phys. Commun. 14, 91 (1978).

384   S. Wilson and D. M. Silver: Comput. Phys. Commun. 17, 47 (1979).

385   B. T. Sutcliffe: in Computational Techniques in Quantum Chemistry
      and Molecular Physics (eds. G. H. F. Diercksen, B. T. Sutcliffe
      and A. Veillard). D. Reidel, Dordrecht-Holland 1975.

386   J. A. Pople: in Computational Methods for Large Molecules and
      Localized States in Solids (eds. F. Herman, A. D. McLean and R.
      K. Nesbet). Plenum Press, New York 1973.

387   M. D. Newton, W. A. Lathan, W. J. Hehre and J. A. Pople: J. Chem.
      Phys. 52, 4064 (1970).

388   J. A. Pople: Acc. Chem. Res. 3, 217 (1970).

389   J. A. Pople: in Modern Theoretical Chemistry, Vol. 4 (ed. H. F.
      Schaefer III). Plenum Press, New York 1977.

390   S. Bell: J. Chem. Phys. 68, 3014 (1978).

391   W. J. Hehre: J. Amer. Chem. Soc. 97, 5308 (1975).

392   P. Rosmus and W. Meyer: J. Chem. Phys. 65, 492 (1976).

393   R. G. Body, D. S. McClure and E. Clementi: J. Chem. Phys. 49,
      4916 (1968).

394   M. Allavena and E. Le Clec'h: J. Mol. Struct. 22, 265 (1974).

395   D. Garton and B. T. Sutcliffe: in Theoretical Chemistry, Vol. 1
      (ed. R. N. Dixon), The Chemical Society, London 1974.

396   P. Pulay: Mol. Phys. 17, 197 (1969).

397   P. Pulay: in Modern Theoretical Chemistry, Vol. 4 (ed. H. F.
      Schaefer III). Plenum Press, New York 1977.

398   D. Poppinger: Chem. Phys. Letters 34, 332 (1975).

399   R. Fletcher and M. J. D. Powell: Computer J. 6, 163 (1963).

400   B. A. Murtagh and R. W. H. Sargent: Computer J. 13, 185 (1970).

401   J. W. McIver, Jr. and A. Komornicki: Chem. Phys. Letters 10, 303
      (1971).

402   J. Pancíř: Collect. Czech. Chem. Commun. 40, 2726 (1975).

403   J. Pancíř: Theoret. Chim. Acta 29, 21 (1973).

404 P. W. Payne: J. Chem. Phys. 65, 1920 (1976).

405a K. Thomsen and P. Swanstrøm: Mol. Phys. 26, 735, 751 (1973).

405b J. D. Goddard, N. C. Handy and H. F. Schaefer III: J. Chem. Phys. 71, 1525 (1979).

405c A. Tachibana, K. Yamashita, T. Yamabe and K. Fukui: Chem. Phys. Letters 59, 255 (1978).

405d S. Kato and K. Morokuma: Chem. Phys. Letters 65, 19 (1979).

406 P. Pulay: Mol. Phys. 18, 473 (1970).

407 H. B. Schlegel, S. Wolfe and F. Bernardi: J. Chem. Phys. 63, 3632 (1975).

408 H. Huber, P. Čársky and R. Zahradník: Theoret. Chim. Acta 41, 217 (1976). H. Huber, J. Pancíř and P. Čársky: Collect. Czech. Chem. Commun. 42, 2767 (1977).

409 M. Dupuis and H. F. King: J. Chem. Phys. 68, 3998 (1978).

410 A. Komornicki, K. Ishida, K. Morokuma, R. Ditchfield and M. Conrad: Chem. Phys. Letters 45, 595 (1977).

411a P. Pulay: Theoret. Chim. Acta 50, 299 (1979).

411b P. Pulay, G. Fogarasi, F. Pang and J. E. Boggs: J. Amer. Chem. Soc. 101, 2550 (1979).

412 H. Huber: Chem. Phys. Letters 62, 95 (1979).

413 J. W. McIver, Jr. and A. Komornicki: J. Amer. Chem. Soc. 94, 2625 (1972); A. Komornicki and J. W. McIver, Jr.: J. Amer. Chem. Soc. 96, 5798 (1974).

414 T. A. Halgren and W. N. Lipscomb: Chem. Phys. Letters 49, 225 (1977).

415 J. G. C. M. van Duijneveldt-van de Rijdt and F. B. van Duijneveldt: J. Mol. Struct. 35, 263 (1976).

416 P. Pulay and W. Meyer: J. Chem. Phys. 57, 3337 (1972).

417 R. H. Schwendeman: J. Chem. Phys. 44, 2115 (1966).

418 J. Pacansky, U. Wahlgren and P. S. Bagus: Theoret. Chim. Acta 41, 301 (1976).

419 C. E. Blom, P. J. Slingerland and C. Altona: Mol. Phys. 31, 1359 (1976).

420 P. Pulay and W. Meyer: Mol. Phys. 27, 473 (1974).

421 P. Botschwina, W. Meyer and A. M. Semkow: Chem. Phys. 15, 25 (1976).

422 G. Das and A. C. Wahl: J. Chem. Phys. 56, 3532 (1972).

423 P. Pulay, W. Meyer and J. E. Boggs: J. Chem. Phys. 68, 5077 (1978).

424 C. E. Blom and C. Altona: Mol. Phys. 31, 1377 (1976).

425 A. Pipano, R. R. Gilman, C. F. Bender and I. Shavitt: Chem. Phys. Letters 4, 583 (1970).

426 R. E. Kari and I. G. Csizmadia: J. Chem. Phys. 56, 4337 (1972).

427 R. M. Stevens: J. Chem. Phys. 55, 1725 (1971).

428 R. M. Stevens: J. Chem. Phys. 61, 2086 (1974).

429 A. Veillard: Theoret. Chim. Acta 18, 21 (1970).

430 T. H. Dunning, Jr. and N. W. Winter: J. Chem. Phys. 63, 1847 (1975).

431 D. Cremer: J. Chem. Phys. 69, 4440 (1978).

432 J. A. Pople: Tetrahedron 30, 1605 (1974).

433 P. W. Payne and L. C. Allen: in Modern Theoretical Chemistry, Vol. 4 (ed. H. F. Schaefer III). Plenum Press, New York 1977.

434 G. Das and A. C. Wahl: Phys. Rev. Letters 24, 440 (1970).

435 V. Kellö, M. Urban, P. Čársky and Z. Slanina: Chem. Phys. Letters 53, 555 (1978).

436 J. A. Pople and J. S. Binkley: Mol. Phys. 29, 599 (1975).

437 P. C. Hariharan and J. A. Pople: Chem. Phys. Letters 16, 217 (1972).

438 J. S. Binkley, J. A. Pople and W. J. Hehre: Chem. Phys. Letters 36, 1 (1975).

439 R. Kosloff, R. D. Levine and R. B. Bernstein: Mol. Phys. 27, 981 (1974).

440 S. Glasstone, K. J. Laidler and H. Eyring: The Theory of Rate Processes. McGraw-Hill, New York 1941.

441 G. J. Janz: Thermodynamic Properties of Organic Compounds. Academic Press, New York 1967.

442 E. B. Wilson, Jr., J. C. Decius and P. C. Cross: Molecular Vibrations. McGraw-Hill, New York 1955.

443 P. Kollman, C. F. Bender and S. Rothenberg: J. Amer. Chem. Soc. 94, 8016 (1972).

444 M. Simonyi and I. Mayer: Acta Chim. Acad. Sci. Hung. 87, 15 (1975).

445 P. Čársky and R. Zahradník: Int. J. Quantum Chem. 16, 243 (1979).

446 P. Čársky: Collect. Czech. Chem. Commun. 44, 3452 (1979).

447 Tables of Interatomic Distances and Configuration in Molecules and Ions (ed. L. E. Sutton). The Chemical Society, London 1958.

448 K. Niblaeus, B. O. Roos and P. E. M. Siegbahn: Chem. Phys. 26, 59 (1977).

449 P. Čársky and R. Zahradník: J. Mol. Struct. 54, 247 (1979).

450 H. S. Johnston: Gas Phase Reaction Rate Theory. Ronald, New York 1966.

451 W. Jakubetz: J. Amer. Chem. Soc. 101, 298 (1979).

452 C. P. Fenimore and G. W. Jones: J. Phys. Chem. 65, 2200 (1961).

453 M. J. Kurylo and R. B. Timmons: J. Chem. Phys. 50, 5076 (1969).

454 M. J. Kurylo, G. A. Hollinden and R. B. Timmons: J. Chem. Phys. 52, 1773 (1970).

455 C. F. Bender, B. J. Garrison and H. F. Schaefer III: J. Chem. Phys. 62, 1188 (1975).

456 W. R. Wadt and N. W. Winter: J. Chem. Phys. 67, 3068 (1977).

457 P. Botschwina and W. Meyer: Chem. Phys. 20, 43 (1977).

458 F. E. Bartoszek, D. M. Manos and J. C. Polanyi: J. Chem. Phys. 69, 933 (1978).

459 T. Koopmans: Physica 1, 104 (1934).

460 H. Basch: J. Electron Spectrosc. and Rel. Phen. 5, 463 (1974).

461 W. G. Richards: Int. J. Mass Spectrom. Ion Phys. 2, 419 (1969).

462a C. L. Pekeris: Phys. Rev. 112, 1649 (1958).

462b G. Verhaegen, J. J. Berger, J. P. Desclaux and C. M. Moser: Chem. Phys. Letters 9, 479 (1971).

463 P. S. Bagus: Phys. Rev. 139A, 619 (1965).

464 P. S. Bagus and H. F. Schaefer III: J. Chem. Phys. 56, 224 (1972).

465 L. S. Cederbaum and W. Domcke: J. Chem. Phys. 66, 5084 (1977).

466 M. M. Coutiére, J. Demuynck and A. Veillard: Theoret. Chim. Acta 27, 281 (1972).

467 M. F. Guest and V. R. Saunders: Mol. Phys. 29, 873 (1975).

468 L. S. Cederbaum, G. Hohlneicher and W. von Niessen: Chem. Phys. Letters 18, 503 (1973).

469 L. S. Cederbaum and W. von Niessen: J. Chem. Phys. 62, 3824 (1975).

470 D. T. Clark, B. J. Cromarty and A. Sgamellotti: J. Electron Spectrosc. and Rel. Phen. 14, 49 (1978) and references therein.

471 M. F. Guest, I. H. Hillier, V. R. Saunders and M. H. Wood: Proc. Roy. Soc. A333, 201 (1973).

472 L. S. Cederbaum and W. Domcke: Advan. Chem. Phys. 36, 205 (1977).

473 L. S. Cederbaum, G. Hohlneicher and W. von Niessen: Mol. Phys. 26, 1405 (1973).

474 L. S. Cederbaum: Theoret. Chim. Acta 31, 239 (1973).

475 J. Paldus and J. Čížek: J. Chem. Phys. 60, 149 (1974).

476 G. D. Purvis and Y. Öhrn: J. Chem. Phys. 60, 4063 (1974).

477 I. Hubač, V. Kvasnička and A. Holubec: Chem. Phys. Letters 23, 381 (1973).

478 I. Hubač and M. Urban: Theoret. Chim. Acta 45, 185 (1977).

479 J. Simons and W. D. Smith: J. Chem. Phys. 58, 4899 (1973).

480 K. M. Griffing and J. Simons: J. Chem. Phys. 64, 3610 (1976).

481 O. W. Day, D. W. Smith and R. C. Morrison: J. Chem. Phys. 62
     115 (1975).

482 D. P. Chong, F. G. Herring and D. McWilliams: J. Chem. Phys.
     61, 958 (1974).

483 B. T. Pickup and O. Goscinski: Mol. Phys. 26, 1013 (1973).

484 M. M. Morrell, R. G. Parr and M. Levy: J. Chem. Phys. 62, 549
     (1975).

485 D. W. Turner, C. Baker, A. D. Baker and C. R. Brundle: Molecular
     Photoelectron Spectroscopy. Wiley-Interscience, London 1970.

486 W. von Niessen, G. H. F. Diercksen and L. S. Cederbaum: Chem.
     Phys. 10, 345 (1975).

487 K. H. Thunemann, R. J. Buenker, S. D. Peyerimhoff and S. K. Shih:
     Chem. Phys. 35, 35 (1978).

488 L. S. Cederbaum and W. Domcke: J. Chem. Phys. 60, 2878 (1974).

489 L. S. Cederbaum and W. Domcke: J. Chem. Phys. 64, 603, 612 (1976).

490 D. P. Chong, F. G. Herring and D. McWilliams: J. Electron
     Spectrosc. and Rel. Phen. 7, 429 (1975).

491 W. Domcke and L. S. Cederbaum: Chem. Phys. 25, 189 (1977).

492 L. S. Cederbaum, W. Domcke, H. Köppel and W. von Niessen: Chem.
     Phys. 26, 169 (1977).

493 H. Köppel, W. Domcke, L. S. Cederbaum and W. von Niessen: J. Chem.
     Phys. 69, 4252 (1978).

494 L. J. Aarons, M. Barber, M. F. Guest, I. H. Hillier and J. H.
     McCartney: Mol. Phys. 26, 1247 (1973).

495 M. H. Wood: Chem. Phys. 5, 471 (1974).

496 G. D. Purvis and Y. Öhrn: J. Chem. Phys. 62, 2045 (1975).

497 S. Svensson, H. Ågren and U. I. Wahlgren: Chem. Phys. Letters 38,
     1 (1976).

498 J. Schirmer, L. S. Cederbaum, W. Domcke and W. von Niessen: Chem.
     Phys. 26, 149 (1977).

499 I. Hubač and V. Kvasnička: Croat. Chim. Acta 49, 677 (1977).

500 M. F. Guest, W. R. Rodwell, T. Darko, I. H. Hillier and J.
     Kendrick: J. Chem. Phys. 66, 5447 (1977).

501 I. B. Ortenburger and P. S. Bagus: Phys. Rev. All, 1501 (1975).

502 H. Ågren, S. Svensson and U. I. Wahlgren: Chem. Phys. Letters 35,
     336 (1975).

503 I. H. Hillier and J. Kendrick: Mol. Phys. 31, 849 (1976).

504 O. Goscinski and H. Siegbahn: Chem. Phys. Letters 48, 568 (1977).

505 D. B. Adams: J. Chem. Soc., Faraday Trans. II 991 (1977).

506 P. S. Bagus and H. F. Schaefer III: J. Chem. Phys. 55, 1474 (1971).

507 W. Domcke, L. S. Cederbaum, J. Schirmer, W. von Niessen and J. P. Maier: J. Electron Spectrosc. and Rel. Phen. 14, 59 (1978).

508 J. Schirmer, L. S. Cederbaum, W. Domcke and W. von Niessen: Chem. Phys. Letters 57, 582 (1978).

509 W. von Niessen, W. Domcke, L. S. Cederbaum and J. Schirmer: J. Chem. Soc., Faraday Trans. II 1550, (1978).

510 L. S. Cederbaum, W. Domcke, J. Schirmer, W. von Niessen, G. H. F. Diercksen and W. P. Kraemer: J. Chem. Phys. 69, 1591 (1978).

511a B. Levy, P. Millie, J. Ridard and J. Vinh: J. Electron Spectrosc. and Rel. Phen. 4, 13 (1974).

511b S. Y. Chu, I. Ozkan and L. Goodman: J. Chem. Phys. 60, 1268 (1974).

512 M. E. Schwartz: in Modern Theoretical Chemistry, Vol. 4 (ed. H. F. Schaefer III). Plenum Press, New York 1977.

513 W. von Niessen, L. S. Cederbaum, W. Domcke and G. H. F. Diercksen: J. Chem. Phys. 66, 4893 (1977).

514 P. W. Deutsch and L. A. Curtiss: Chem. Phys. Letters 39, 588 (1976).

515 H. Ågren, S. Svensson and U. I. Wahlgren: Int. J. Quantum Chem. 11, 317 (1977).

516 W. von Niessen, G. H. F. Diercksen and L. S. Cederbaum: J. Chem. Phys. 67, 4124 (1977).

517 J. O. Hirschfelder, C. F. Curtiss and R. B. Bird: Molecular Theory of Gases and Liquids. Wiley, New York 1954.

518 H. Margenau and N. R. Kestner: Theory of Intermolecular Forces. Pergamon, Oxford 1971.

519 P. Hobza and R. Zahradník: Weak Intermolecular Interaction in Chemistry and Biology. Elsevier, Amsterdam 1980.

520 A. T. Amos and R. J. Crispin: in Theoretical Chemistry, Vol. 2 (eds. H. Eyring and D. Henderson). Academic Press, New York 1976.

521 M. Dreyfus and A. Pullman: Theoret. Chim. Acta 19, 20 (1970).

522 E. Kochanski: J. Chem. Phys. 58, 5823 (1973).

523 R. Ahlrichs: Theoret. Chim. Acta 41, 7 (1976).

524 J. O. Hirschfelder: Int. J. Quantum Chem. S4, 257 (1971).

525 J. N. Murrell, M. Randić and D. R. Williams: Proc. Roy. Soc. A284, 566 (1965).

526 J. N. Murrell and G. Shaw: J. Chem. Phys. 46, 1768 (1967).

527 P. Claverie: Int. J. Quantum Chem. 5, 273 (1971).

528 J. P. Daudey, P. Claverie and J. P. Malrieu: Int. J. Quantum Chem. 8, 1 (1974).

529 W. Kutzelnigg: Faraday Discuss. 62, 185 (1977).

530 B. Jeziorski and W. Koŕos: Int. J. Quantum Chem. 12 Suppl.1, 91
    (1977). B. Jeziorski, K. Szalewicz and G. Chaŕasiński: Int. J.
    Quantum Chem. 14, 271 (1978).

531 W. Kutzelnigg and F. Maeder: Chem. Phys. 32, 451 (1978).

532 F. Maeder and W. Kutzelnigg: Chem. Phys. 32, 457 (1978).

533 P. A. Kollman and L. C. Allen: Theoret. Chim. Acta 18, 399 (1970).

534 K. Morokuma: J. Chem. Phys. 55, 1236 (1971).

535 K. Kitaura and K. Morokuma: Int. J. Quantum Chem. 10, 325 (1976).

536 S. Iwata and K. Morokuma: J. Amer. Chem. Soc. 95, 7563 (1973).

537a A. Beyer: Dissertation, University of Vienna, 1976.

537b P. Schuster, H. Lischka and A. Beyer: in Progress in Theoretical
    Organic Chemistry, Vol. 2 (ed. I. G. Csizmadia). Elsevier,
    Amsterdam 1977.

538 H. Umeyama and K. Morokuma: J. Amer. Chem. Soc. 99, 1316 (1977).

539 H. Popkie, H. Kistenmacher and E. Clementi: J. Chem. Phys. 59,
    1325 (1973).

540 P. A. Kollman and L. C. Allen: Chem. Rev. 72, 283 (1972).

541 L. C. Allen: J. Amer. Chem. Soc. 97, 6921 (1975).

542 P. Schuster: in The Hydrogen Bond—Recent Developments in Theory
    and Experiments, Vol. 1 (eds. P. Schuster, G. Zundel and C.
    Sandorfy). North Holland, Amsterdam 1976.

543 S. F. Boys and F. Bernardi: Mol. Phys. 19, 553 (1970).

544 E. Clementi: J. Chem. Phys. 46, 3851 (1967).

545 A. Johansson, P. Kollman and S. Rothenberg: Theoret. Chim. Acta
    29, 167 (1973).

546 M. Urban and P. Hobza: Theoret. Chim. Acta 36, 215 (1975).

547 N. S. Ostlund and D. L. Merrifield: Chem. Phys. Letters 39, 612
    (1976).

548 M. Urban, S. Pavlík, V. Kellö and J. Mardiaková: Collect. Czech.
    Chem. Commun. 40, 587 (1975).

549 M. D. Newton and S. Ehrenson: J. Amer. Chem. Soc. 93, 4971 (1971).

550 J. P. Daudey, J. P. Malrieu and O. Rojas: Int. J. Quantum Chem.
    8, 17 (1974).

551 J. P. Daudey: Int. J. Quantum Chem. 8, 29 (1974).

552a B. Liu and A. D. McLean: J. Chem. Phys. 59, 4557 (1973).

552b P. D. Dacre: Chem. Phys. Letters 50, 147 (1977).

552c S. L. Price and A. J. Stone: Chem. Phys. Letters 65, 127 (1979).

553 M. Urban, S. Hrivnáková and P. Hobza: unpublished.

554 P. E. Cade and W. M. Huo: J. Chem. Phys. 45, 1063 (1966).

555 A. D. Buckingham: Quart. Rev. 13, 183 (1959).

556 J. G. C. M. van Duijneveldt-van de Rijdt and F. B. van Duijneveldt:
J. Amer. Chem. Soc. 93, 5644 (1971).

557 M. Mezei and E. S. Campbell: Theoret. Chim. Acta 43, 227 (1977).

558 P. Schuster, W. Marius, A. Pullman and H. Berthod: Theoret. Chim.
Acta 40, 323 (1975).

559 V. Staemmler: Chem. Phys. 7, 17 (1975).

560 B. J. Garrison, W. A. Lester, Jr. and H. F. Schaefer III: J. Chem.
Phys. 63, 1449 (1975).

561 P. E. S. Wormer and A. van der Avoird: J. Chem. Phys. 62, 3326
(1975).

562 F. Mulder, M. van Hemert, P. E. S. Wormer and A. van der Avoird:
Theoret. Chim. Acta 46, 39 (1977).

563 D. Perahia, A. Pullman and B. Pullman: Theoret. Chim. Acta 42, 23
(1976).

564 D. Perahia, A. Pullman and B. Pullman: Theoret. Chim. Acta 43,
207 (1977).

565 E. Scrocco and J. Tomasi: Top. Curr. Chem. 42, 95 (1973).

566 P. Kollman: J. Amer. Chem. Soc. 99, 4875 (1977).

567 M. Trenary, H. F. Schaefer III and P. Kollman: J. Amer. Chem.
Soc. 99, 3885 (1977).

568 J. Prissette and E. Kochanski: Chem. Phys. Letters 47, 391 (1977).

569 M. Jaszuński, E. Kochanski and P. Siegbahn: Mol. Phys. 33, 139
(1977).

570 M. Trenary, H. F. Schaefer III and P. A. Kollman: J. Chem. Phys.
68, 4047 (1978).

571 V. A. Nicely and J. L. Dye: J. Chem. Phys. 52, 4795 (1970).

572 J. E. Del Bene: J. Chem. Phys. 62, 666 (1975).

573 H. Umeyama, K. Morokuma and S. Yamabe: J. Amer. Chem. Soc. 99, 330
(1977).

574 E. Kochanski, B. Roos, P. Siegbahn and M. H. Wood: Theoret. Chim.
Acta 32, 151 (1973).

575 G. Das and A. C. Wahl: Phys. Rev. A4, 825 (1971).

576 D. R. Salahub: Mol. Phys. 28, 243 (1974).

577 P. Hobza, P. Čársky and R. Zahradník: Int. J. Quantum Chem. 16,
257 (1979).

578 J. D. Dill, L. C. Allen, W. C. Topp and J. A. Pople: J. Amer.
Chem. Soc. 97, 7220 (1975).

579 P. Kollman and S. Rothenberg: J. Amer. Chem. Soc. 99, 1333 (1977).

580 R. L. Woodin, F. A. Houle and W. A. Goddard III: Chem. Phys. 14,
461 (1976).

581 A. Pullman: in The World of Quantum Chemistry (eds. R. Daudel and B. Pullman). D. Reidel, Dordrecht 1974.

582 P. Schuster, W. Jakubetz and W. Marius: Topics Curr. Chem. 60, 1 (1975).

583 A. Pullman and B. Pullman: Quart. Rev. Biophys. 7, 505 (1975).

584 E. Clementi: Determination of Liquid Water Structure. Coordination Numbers of Ions and Solvation for Biological Molecules. Lecture Notes in Chemistry, Vol. 2. Springer-Verlag, Berlin 1976.

585 P. Kebarle: Ann. Rev. Phys. Chem. 28, 445 (1977).

586 P. Hobza and R. Zahradník: Topics Curr. Chem., in press.

587 I. Džidić and P. Kebarle: J. Phys. Chem. 74, 1466 (1970).

588 B. O. Roos, W. P. Kraemer and G. H. F. Diercksen: Theoret. Chim. Acta 42, 77 (1976).

589 J. A. Barker and R. O. Watts: Chem. Phys. Letters 3, 144 (1969).

590 C. N. Sarkisov, V. G. Dashevsky and G. G. Malenkov: Mol. Phys. 27, 1249 (1974).

591 G. C. Lie and E. Clementi: J. Chem. Phys. 62, 2195 (1975).

592 O. Matsuoka, E. Clementi and M. Yoshimine: J. Chem. Phys. 64, 1351 (1976).

593 G. C. Lie, E. Clementi and M. Yoshimine: J. Chem. Phys. 64, 2314 (1976).

594 S. Swaminathan and D. L. Beveridge: J. Amer. Chem. Soc. 99, 8392 (1977).

595 H. Kistenmacher, G. C. Lie, H. Popkie and E. Clementi: J. Chem. Phys. 61, 546 (1974).

596 R. O. Watts, E. Clementi and J. Fromm: J. Chem. Phys. 61, 2550 (1974).

597 J. Fromm, E. Clementi and R. O. Watts: J. Chem. Phys. 62, 1388 (1975).

598 H. Kistenmacher, H. Popkie and E. Clementi: J. Chem. Phys. 58, 5627 (1973).

599 H. Kistenmacher, H. Popkie and E. Clementi: J. Chem. Phys. 61, 799 (1974).

600 E. Clementi, R. Barsotti, J. Fromm and R. O. Watts: Theoret. Chim. Acta 43, 101 (1976); E. Clementi and R. Barsotti: Chem. Phys. Letters 59, 21 (1978).

601 E. Clementi, F. Cavallone and R. Scordamaglia: J. Amer. Chem. Soc. 99, 5531 (1977).

602 R. Scordamaglia, F. Cavallone and E. Clementi: J. Amer. Chem. Soc. 99, 5545 (1977).

603 G. Bolis and E. Clementi: J. Amer. Chem. Soc. 99, 5550 (1977).

604 G. Ranghino and E. Clementi: Gazz. Chim. Ital. 108, 157 (1978).

605 S. Romano and E. Clementi: Gazz. Chim. Ital. 108, 319 (1978).

606 E. Clementi and G. Corongiu: Chem. Phys. Letters 60, 175 (1979);
Int. J. Quantum Chem. 16, 897 (1979).

607 E. Clementi, G. Ranghino and R. Scordamaglia: Chem. Phys. Letters
49, 218 (1977).

608 J. Del Bene and J. A. Pople: J. Chem. Phys. 52, 4858 (1970).

609 D. Hankins, J. W. Moskowitz and F. H. Stillinger: J. Chem. Phys.
53, 4544 (1970).

610 B. R. Lentz and H. A. Scheraga: J. Chem. Phys. 58, 5296 (1973);
61, 3493 (1974).

611 H. Veillard, J. Demuynck and A. Veillard: Chem. Phys. Letters 33,
221 (1975).

612 P. A. Kollman and I. D. Kuntz: J. Amer. Chem. Soc. 94, 9236 (1972).

613 W. P. Kraemer and G. H. F. Diercksen: Theoret. Chim. Acta 27, 265
(1972).

614 W. P. Kraemer and G. H. F. Diercksen: Theoret. Chim. Acta 23, 393
(1972).

615 P. A. Kollman and I. D. Kuntz: J. Amer. Chem. Soc. 96, 4766 (1974).

616 A. Pullman, H. Berthod and N. Gresh: Chem. Phys. Letters 33, 11
(1975).

617 B. Pullman: in Environmental Effects on Molecular Structure and
Properties. The Jerusalem Symposia on Quantum Chemistry and Biochemistry, Vol. 8 (ed. B. Pullman). Reidel, Dordrecht 1976.

618 J. E. Del Bene: J. Amer. Chem. Soc. 96, 5643 (1974).

619 S. Iwata and K. Morokuma: J. Amer. Chem. Soc. 97, 966 (1975).

620 L. A. Curtiss and J. A. Pople: J. Mol. Spectrosc. 55, 1 (1975).

621 J. Sadlej and A. J. Sadlej: Faraday Discuss. 64, 112 (1978).

622 J. Sadlej and A. J. Sadlej: Acta Phys. Pol. A53, 747 (1978).

623 G. H. F. Diercksen, W. von Niessen and W. P. Kraemer: Theoret.
Chim. Acta 31, 205 (1973).

624 R. Janoschek, E. G. Weidemann, H. Pfeiffer and G. Zundel: J.
Amer. Chem. Soc. 94, 2387 (1972).

625 G. Zundel: Hydration and Intermolecular Interaction – Infrared
Investigation with Polyelectrolyte Membranes. Academic Press,
New York 1969.

626 G. Zundel: in The Hydrogen Bond-Recent Developments in Theory
and Experiments, Vol. 2 (eds. P. Schuster, G. Zundel and C.
Sandorfy). North-Holland, Amsterdam 1976.

627 I. Černušák and M. Urban: Collect. Czech. Chem. Commun. 43, 1956
(1978).

628 P. O. Löwdin: Rev. Mod. Phys. 35, 724 (1963).

629 R. Janoschek: Theoret. Chim. Acta 32, 49 (1973).

630 J. J. Delpuech, G. Serratrice, A. Strich and A. Veillard: Mol. Phys. 29, 849 (1975).

631 P. Kollman and I. Kuntz: J. Amer. Chem. Soc. 98, 6820 (1976).

632 J. O. Noell and K. Morokuma: J. Phys. Chem. 80, 2675 (1976).

633 P. Merlet, S. D. Peyerimhoff and R. J. Buenker: J. Amer. Chem. Soc. 94, 8301 (1972).

634 W. Meyer, W. Jakubetz and P. Schuster: Chem. Phys. Letters 21, 97 (1973).

635 A. Støgard, A. Strich, J. Almlöf and B. Roos: Chem. Phys. 8, 405 (1975).

636 S. Ray: Chem. Phys. Letters 11, 573 (1971).

637 A. Johansson and M. Jäntti: Finn. Chem. Letters 46 (1976).

638 S. Yamabe, S. Kato, H. Fujimoto and K. Fukui: Theoret. Chim. Acta 30, 327 (1973).

639 J. Almlöf, J. Lindgren and J. Tegenfeldt: J. Mol. Struct. 14, 427 (1972); J. Almlöf and U. Wahlgren: Theoret. Chim. Acta 28, 161 (1973).

640 J. Hylton McCreery, R. E. Christoffersen and G. G. Hall: J. Amer. Chem. Soc. 98, 7198 (1976).

641 J. O. Noell and K. Morokuma: Chem. Phys. Letters 36, 465 (1975).

642 D. M. Hayes and P. A. Kollman: J. Amer. Chem. Soc. 98, 3335, 7811 (1976).

643 G. Klopman: Chem. Phys. Letters 1, 200 (1967).

644 J. Hylton, R. E. Christoffersen and G. G. Hall: Chem. Phys. Letters 24, 501 (1974); J. Hylton McCreery, R. E. Christoffersen and G. G. Hall: J. Amer. Chem. Soc. 98, 7191 (1976).

645 O. Tapia and O. Goscinski: Mol. Phys. 29, 1653 (1975).

646 O. Tapia: Theoret. Chim. Acta 47, 157 (1978).

647 H. A. Germer, Jr.: Theoret. Chim. Acta 34, 145 (1974); 35, 273 (1974).

648 F. Birnstock, H. J. Hofmann and H. J. Köhler: Theoret. Chim. Acta 42, 311 (1976).

649 P. Claverie, J. P. Daudey, J. Langlet, B. Pullman, D. Piazzola and M. J. Huron: J. Phys. Chem. 82, 405 (1978).

650 D. L. Beveridge and G. W. Schnuelle: J. Phys. Chem. 78, 2064 (1974).

651 G. W. Schnuelle and D. L. Beveridge: J. Phys. Chem. 79, 2566 (1975).

652 J. L. Burch, K. S. Raghuveer and R. E. Christoffersen: in Environmental Effects on Molecular Structure and Properties. The Jerusalem Symposia on Quantum Chemistry and Biochemistry, Vol. 8 (ed. B. Pullman). D. Reidel, Dordrecht 1976.

653 E. Clementi and H. Popkie: Chem. Phys. Letters 20, 1 (1973).

654 J. Almlöf: Int. J. Quantum Chem. 8, 915 (1974).

655 H. E. Popkie, W. S. Koski and J. J. Kaufman: J. Amer. Chem. Soc. 98, 1342 (1976).

656 H. E. Popkie and J. J. Kaufman: Int. J. Quantum Chem. 10, 569 (1976).

657 A. Veillard and J. Demuynck: in Modern Theoretical Chemistry, Vol. 4. (ed. H. F. Schaefer III). Plenum Press, New York 1977.

658 P. S. Bagus, U. I. Wahlgren and J. Almlöf: J. Chem. Phys. 64, 2324 (1976).

659 K. Ohno: Int. J. Quantum Chem. 12, Suppl. 1, 119 (1977).

660 H. Kashiwagi, T. Takada, S. Obara, E. Miyoshi and K. Ohno: Int. J. Quantum Chem. 14, 13 (1978).

661 R. E. Christoffersen: Int. J. Quantum Chem. 16, 573 (1979).

662 G. H. F. Diercksen and W. P. Kraemer: in Computational Techniques in Quantum Chemistry and Molecular Physics (eds. G. H. F. Diercksen, B. T. Sutcliffe and A. Veillard). D. Reidel, Dordrecht 1975.

663 H. F. Schaefer III and W. H. Miller: Computers & Chemistry 1, 85 (1976).

664 R. A. Sparks: Int. J. Quantum Chem. S12, 191 (1978).

665 R. R. Lucchese, B. R. Brooks, J. H. Meadows, W. C. Swope and H. F. Schaefer III: J. Computational Phys. 26, 243 (1978).

666 B. R. Brooks and H. F. Schaefer III: Int. J. Quantum Chem. 14, 603 (1978).

667 R. R. Lucchese, M. P. Conrad and H. F. Schaefer III: J. Chem. Phys. 68, 5292 (1978).

668 L. R. Kahn, P. Baybutt and D. G. Truhlar: J. Chem. Phys. 65, 3826 (1976).

669 D. McWilliams and S. Huzinaga: J. Chem. Phys. 63, 4678 (1975).

670 H. E. Popkie and J. J. Kaufman: Int. J. Quantum Chem. S11, 433 (1977).

671 C. F. Melius, B. D. Olafson and W. A. Goddard III: Chem. Phys. Letters 28, 457 (1974).

672 P. Coffey, C. S. Ewig and J. R. Van Wazer: J. Amer. Chem. Soc. 97, 1656 (1975).

673 R. Osman and H. Weinstein: Chem. Phys. Letters 49, 69 (1977).

674 S. Topiol, M. A. Ratner and J. W. Moskowitz: Chem. Phys. 20, 1 (1977).

675 S. Huzinaga and M. Yoshimine: J. Chem. Phys. 68, 4486 (1978).

676 A. Pullman, N. Gresh, J. P. Daudey and J. W. Moskowitz: Int. J. Quantum Chem. S11, 501 (1977).

677 W. Marius and P. Schuster: Theoret. Chim. Acta 42, 5 (1976).

678 Ch. Teichteil, J. P. Malrieu and J. C. Barthelat: Mol. Phys. 33, 181 (1977).

679 W. H. E. Schwarz: Communicated on the 4. Arbeitstagung für Theoretische Chemie in Mariapfar. Austria 1978.

680 T. C. Chang, P. Habitz and W. H. E. Schwarz: Theoret. Chim. Acta 44, 61 (1977).

681 Y. S. Lee, W. C. Ermler and K. S. Pitzer: J. Chem. Phys. 67, 5861 (1977).

682 L. R. Kahn, P. J. Hay and R. D. Cowan: J. Chem. Phys. 68, 2386 (1978).

683 G. Das and A. C. Wahl: J. Chem. Phys. 69, 53 (1978).

684 P. Hafner and W. H. E. Schwarz: J. Phys. B 11, 217 (1978).

685 P. J. Hay, W. R. Wadt, L. R. Kahn, R. C. Raffenetti and D. H. Phillips: J. Chem. Phys. 71, 1767 (1979) and references therein.

686 J. Paldus: in Theoretical Chemistry: Advances and Perspectives, Vol. 2 (eds. H. Eyring and D. J. Henderson). Academic Press, New York 1976.

687 J. Paldus: in Electrons in Finite and Infinite Structures (eds. P. Phariseau and L. Scheire). Plenum Press, New York 1977.

688 I. Shavitt: Int. J. Quantum Chem. S11, 131 (1977); S12, 5 (1978).

689 B. R. Brooks and H. F. Schaefer III: J. Chem. Phys. 70, 5092 (1979).

690 P. E. M. Siegbahn: J. Chem. Phys. 70, 5391 (1979).

691 R. F. Hausman, Jr., S. D. Bloom and C. F. Bender: Chem. Phys. Letters 32, 483 (1975).

692 R. F. Hausman, Jr. and C. F. Bender: in Modern Theoretical Chemistry, Vol. 3 (ed. H. F. Schaefer III). Plenum Press, New York 1977.

693 E. R. Davidson and C. F. Bender: Chem. Phys. Letters 59, 369 (1978).

694 N. H. F. Beebe and J. Linderberg: Int. J. Quantum Chem. 12, 683 (1977).

695 P. E. M. Siegbahn: Chem. Phys. 25, 197 (1977).

696 W. Meyer: J. Chem. Phys. 64, 2901 (1976).

697 C. E. Dykstra, H. F. Schaefer III and W. Meyer: J. Chem. Phys. 65, 2740 (1976).

698 E. Clementi: J. Chem. Phys. 47, 2323 (1967).

699 B. S. Ault and G. C. Pimentel: J. Phys. Chem. 77, 1649 (1973).

700 N. C. Baird: Pure Appl. Chem. 49, 223 (1977).

701 P. Hennig, W. P. Kraemer and G. H. F. Diercksen: A Compilation of Theoretical Spectroscopic Constants and Rotational-Vibrational Transition Frequencies for the Isoelectronic Series of Linear Triatomic Molecules HCN, HNC, HCO^+, HOC^+, HNN^+ Obtained from Ab Initio Calculated Energy Hypersurface. Max-Planck-Institut für Physik und Astrophysik, Munich 1977.

702 H. F. Schaefer III: Accounts Chem. Res. 10, 287 (1977).

703 D. H. Whiffen: Pure Appl. Chem. 50, 75 (1978).

704 E. R. Cohen and B. N. Taylor: J. Phys. Chem. Ref. Data 2, 663 (1973).

705 J. S. Binkley, R. A. Whiteside, P. C. Hariharan, R. Seeger, J. A. Pople, W. J. Hehre and M. D. Newton: GAUSSIAN 76, QCPE 368. Indiana University, Bloomington.

706 P. D. Mallinson: Molecular Wave Functions by Gaussian 70. User Guide. University of London, Computer Center, London 1976.

707 T. D. Metzgar and J. E. Bloor: POLYATOM Version II (IBM-360). QCPE 238, Indiana University, Bloomington.

708 D. Goutier, R. Macaulay and A. J. Duke: PHANTOM, Ab initio Quantum Chemical Programs for CDC 6000 and 7000 Series Computers. QCPE 241, Indiana University, Bloomington.

709 M. Dupuis, J. Rys and H. F. King: HONDO 76, QCPE 336 and 338. Indiana University, Bloomington.

Subject Index

THEORETICA CHIMICA ACTA

an International Journal
of Theoretical Chemistry

ISSN 0040-5744   TitleNo.214

Edenda curat: Hermann Hartmann, Mainz

Adiuvantibus: C.J.Ballhausen, København; R.D.Brown,
Clayton; K.Fukui, Kyoto; R.Gleiter, Heidelberg;
E.A.Halevi, Haifa; G.G.Hall, Nottingham; E.Heilbronner,
Basel; J.Jortner, Tel-Aviv; M.Kotani, Tokyo; J.Koutecký,
Berlin; A.Neckel, Wien; E.E.Nikitin, Moskwa; R.G.
Pearson, Santa Barbara; B.Pullmann, Paris; B.Rånby, Stock-
holm; K.Ruedenberg, Ames; C.Sandorfy, Montreal;
M.Simonetta, Milano; O.Sinanoğlu, NewHaven;
R.Zahradník, Praha

Today, theory and experiment are inseparably bound. Every
chemical experiment is preceded by reflection and careful
consideration, and the results are interpreted according to
chemical theories and perceptions.

The editors of **Theoretica Chimica Acta** therefore wish to
emphasize the wide-ranging program reflected in the policy
of their journal:

"**Theoretica Chimica Acta** accepts manuscripts in which the
relationships between individual chemical and physical
phenomena are investigated. In addition, experimental
research that presents new theoretical viewpoints is desired."

Theoretica Chimica Acta offers experimental chemists increased space for the publication of discussion of the goals of
their work, the significance of their findings, and the
concepts on which their experimental work is based. Such
discussions contribute significantly to mutual understanding between theoreticians and experimentalists and stimulate
both new reflections and further experiments.

Springer
International

Subscription Information and/or sample copies upon request.
Please send your order or request to your bookseller
or directly to:
Springer-Verlag, Journal Promotion Department,
P.O.Box 105280, D-6900 Heidelberg, FRG